# CREATIVITY BOOKS BY ED GLASSMAN

Team Creativity At Work-I. You Do Want To Be More Successful Than Your Competition, Don't You? (2010).

Team Creativity At Work-II. Brainstorming Isn't Creative Enough Anymore. (2010).

The Creativity Factor: Unlocking the Potential of Your Team (1991) San Diego, CA: Pfeiffer Books of University Associates.

For Presidents Only: Unlocking the Creative Potential of Your Management Team. (1990) NYC: The Presidents Association of the American Management Association.

Creativity Handbook: A Practical Guide To Shift Paradigms And Improve Creative Thinking At Work (1996) A 250 page book written for his creativity workshops.

# R&D Creativity & Innovation Handbook:

### A Practical Guide To Improve Creative Thinking and Innovation Success At Work

by

Edward Glassman, Ph.D.

Professor Emeritus

University of North Carolina

Chapel Hill

Former President, The Creativity College®,

A Division of Leadership Consulting Services, Inc.

Former Professor and Program Head
The Program For Team Effectiveness And Creativity
The University of North Carolina At Chapel Hill

*Imagine R&D TEAMS Shifting Paradigms IN A Creative Climate using R&D CREATIVITY TOOLS and Techniques DURING Creativity & Innovation Meetings THINKING CREATIVELY and Solving R&D Problems INNOVATIVELY…*

Brainstorming Isn't Creative Enough Anymore

📖

CreateSpace

**r-and-d-creativity-innovation.com**

# CREATIVITY BOOKS BY ED GLASSMAN

- Team Creativity At Work I. You Do Want To Be More Successful Than Your Competition, Don't You? (2010).

- Team Creativity At Work II. Brainstorming Isn't Creative Enough Anymore. (2010).

- The Creativity Factor: Unlocking the Potential of Your Team (1991) San Diego, CA: Pfeiffer Books of University Associates.

- For Presidents Only: Unlocking the Creative Potential of Your Management Team. (1991) NYC: The Presidents Association of the American Management Association.

- Creativity Handbook: A Practical Guide To Shift Paradigms And Improve Creative Thinking At Work (1996) A 250 page book written to use in his creativity & innovation meetings & workshops.

-------------------------------------------------------------------

# PREFACE by Dr. Dave Tanner (now retired) *

## Former Director, DuPont Center for Creative Thinking & Innovation,

To maintain global leadership it's essential that we out think and out perform our competitors. Our natural tendency is to tackle problems and opportunities building from our long standing experience base. This type of linear thinking is important and a logical starting point. It is particularly valuable in team efforts where team members have a diverse experience base.

However, to accomplish revolutionary as well as evolutionary change often requires that we step outside our normal experience base and engage in non linear thinking. For most people this is difficult! That's where the technology of creative thinking comes in. Many tools exist that force us outside our box of conventional thinking. That's what this book by Ed Glassman is about.

According to Ed, who long ago was trained as a geneticist, the differences people display in ability to think creatively is probably not inherited. Published research quoted in Chapter 4 of this book using identical and fraternal twins supports this view. Creative thinking is a learnable skill. My experience in over 20 years as a director of research and development also reinforces this view.

We have many examples where step-change advances with bottom line impact have derived from an individual team practicing creative thinking tools of the type described in this book.

Ed Glassman has been a creative thinking-enhancing specialist consulting for DuPont since 1983. He and others effectively teach the tools of creative thinking in seminars and as part of workshops designed to tackle practical problems.

In this book, Ed describes many of these tools. He also touches on the creative climate which is necessary if creative ideas are to flourish and be implemented. Books like this are welcome additions to our library of important works in the newly formed DuPont Center for Creative thinking and Innovation.

-----------------------------------------------------------------

*Dave Tanner, Ph.D., is a creative DuPont scientist who holds 33 patents from his early years in research. He started at DuPont in 1954 as a research chemist. Before becoming the founding Director of the Dupont Center for Creativity & Innovation, he held management positions at several DuPont sites, including laboratory director, fibers strategic planning manager, and director of industrial fibers R&D.

# THIS BOOK IS GRATEFULLY DEDICATED TO:

- My friend, Vicki Bradley, who creates.
- My parents, who, in their own way, encouraged me to write this book when I was a child.
- My four marvelous daughters, Ellen, Marjorie, Lyn, and Susan, and their splendid children, my grandchildren, Annie, Charlotte, Deborah, Dylan, Joshua, Julia, Nick, Rebecca, Sarah, and Trey.

-----------------------------------------------------------------

# ACKNOWLEDGEMENTS

My creative friend, Vicki Bradley, helped me in the presentation of numerous workshops. Much of the material in this book came from these workshops and the countless discussions she and I have had about creative thinking, especially hers.

In 1983, I was a Visiting Fellow at The Center For Creative Leadership in Greensboro, NC. Many people there helped to sow the seeds of this book. Some of these are: David Campbell, who inspired; Bob Dorn, who befriended; Bob Bailey, who clarified; Bill Drath, who helped me write better; and David DeVries, who advised with wisdom. I am grateful to all of them.

Many managers and professionals in corporations read and gave me comments on various versions of this book; to them I am very grateful. They are too numerous to mention but their input appears on many of these pages.

Finally, I want to thank the thousands of people who attended my workshops and kindly gave me feedback on what worked and what needed improvement. Without their input, I could not have continually improved and upgraded the workshop material that led to this book. I will always value and appreciate knowing them.

# ABOUT THE AUTHOR

EDWARD GLASSMAN, Ph.D. was the president of The Creativity College®, a division of Leadership Consulting Services, Inc. and Professor of the University of North Carolina at Chapel Hill (now retired), where he founded & headed The Program For Team Effectiveness And Creativity.

A Guggenheim Foundation Fellow and a Visiting Professor at Stanford University (1968-69) and a Visiting Fellow at The Center For Creative Leadership in Greensboro, North Carolina (1986), his biography appears in "Who's Who In America" and "Who's Who in the World."

Professor Glassman wrote many newspaper columns for local newspapers on such topics as: "Business Creativity" and "Creativity at Work."

His articles on creative thinking and on team excellence appeared in Supervisory Management; R&D Management; Intrepreneurial Excellence; The Female Executive; Laboratory Management; Management Solutions; and The President.

Glassman led scores of Creativity & Innovation Meetings and workshops for large and small organizations, including DuPont; Amoco Chemical; IBM; Texaco; Ciba-Geigy; Hoechst-Celanese; Milliken; Federal-Mogul; Calreco/Carnation; A.T.&T. Bell Laboratories; Standard Products; Eastman Chemical; Thetford; Lucas Engineering (UK); and others.

He wrote books on creativity and team excellence including:

• "Team Creativity-I, You Do Want To Be More Successful Than Your Competition, Don't You?" (2010).

• "Team Creativity-II. Brainstorming Isn't Creative Enough Anymore." (2010).

• "For Presidents Only: Unlocking The Creative Potential Of Your Management Team." (1990) The Presidents Association of The American Management Association, NY.

• "The Creativity Factor: Unlocking the Potential of Your Team," (1991) Pfeiffer & Company.

• "Creativity Handbook: A Practical Guide to Paradigm Shifts and Creative Thinking at Work," (1996) A 250 page manual used in his creativity workshops & meetings.

Born March 18, 1929, in NYC and graduated from Stuyvesant High School in 1946, he received his Ph.D. in 1955 from the Biology Department, the Johns Hopkins University.

He was a Professor (now retired) in the University of North Carolina; Chapel Hill, NC, from 1960 to 1989, and published over 100 research articles on biochemistry, genetics, neuroscience, alcohol, teaching, and creativity.

He served on the Editorial Boards of 'Neurochemical Research' (1975-1978); 'Behavioral Biology' (1971-1976); 'Pharmacology, Biochemistry, and Behavior' (1973-1988); and 'Behavior Genetics' (1970-1971).

# PERTINENT ARTICLES AND BOOKS BY EDWARD GLASSMAN

(1982) A leadership skills program for scientist-supervisors. Laboratory Management (Sept), p. 46-49.

(1993) Your leadership style: How to capitalize on your subordinates' perceptions of you. Executive Female 6(5): 29-33 (Sept-October).

(1986) Creativity for greater productivity. Interview by Boardroom Reports (March 15), p. 3-4.

(1986) Leadership's styles effect on the creativity of employees. Management Solutions 31: 18-25.

(1986) Managing for creative thinking: Back to basics in R&D. R&D Management 16:175-183.

(1986) Habits that need changing. Intrapreneurial Excellence (June), p.4.

(1987) Your Leadership Style (in "Leadership,"edited by A. D. Timpe. Volume 3 of 'The Arts and Science of Business Management'" published by Facts On File Publications, p 117-121.

(1988) Are your workers as creative as they could be? Management Solutions (Oct) p. 29-31.

(1989) Creative problem solving. Supervisory Management (January) p. 21-26.

(1989) Creative problem solving: Habits that need changing. Supervisory Management (Feb) p. 8-12.

(1989) Some techniques for problem solving. Supervisory Management (March) p.14-18.

(1989) Creative problem solving: Your role as leader. Supervisory Management,(April) p. 37-42.

(1989) Creative team building without a consultant. The President (published by the American Management Association)

25 (No 8): Sept, p. 4.

(1990) "For Presidents Only: Unlocking the Creative Potential of Your Management Team." New York: Presidents Association, American Management Association.

(1990) Understanding and supervising low conformers. Supervisory Management 35 (November), p 10.

(1990) Leadership's styles effect on the creative thinking of employees. Pins and Needles National Productivity Board of Singapore. Issue 5:34-38.

(1991) "The Creativity Factor: Unlocking the Potential of Your Team." San Diego, CA: Pfeiffer Books of University Associates.

(1991) Selling your ideas to management. Supervisory Management 39: (October) p 9.

(1991) Self directed team building without a consultant. Supervisory Management. (March) p. 6.

(1996) Creativity Handbook: A Practical Guide To Shift Paradigms And Improve Creative Thinking At Work, a 250 page book written for his creativity & innovation meetings & workshops.

(2010) Team Creativity At Work I. You Do Want To Be More Successful Than Your Competition, Don't You.

(2010) Team Creativity At Work II. Brainstorming Isn't Creative Enough Anymore.

# TABLE OF CONTENTS

## PART 3
## TECHNIQUES TO SHIFT PARADIGMS & SOLVE PROBLEMS CREATIVELY

## PART 4
## TECHNIQUES FOR CREATIVE R&D MEETINGS.

## PART 7
## CONVERT IDEAS INTO PROFITABLE INNOVATIONS IN R&D

## PART 8
## APPENDICES

---

> # • INTRODUCTION •
> ## The High Quality Innovation Solution <u>vs</u> The Quick Fix

When innovation succeeds, the company flourishes. For innovation to succeed, solving innovation problems must capture the creative thinking of all people. Achieve high-quality solutions to innovation problems in R&D the new-fashioned way: apply modern creative thinking techniques. These techniques provide the thinking skills necessary for innovation at work. High quality solutions appear only after intentional effort to generate and select from a range of perspectives and ideas that create new innovation possibilities.

Quality innovation solutions depend on a variety of thoughts, experience, training, expertise, interactions among people, and the subconscious mind, all coming together to produce a bright outcome. Unless this process includes discipline and conformity to creativity guidelines, the innovation effort produces chaos with an occasional idea, often a **quick-fix** solution.

Creative thinking techniques change that, providing the focus and clarity that lead individuals and teams to create quality innovation solutions that gratify and delight. Only careful, deliberative use of creativity avoids the quick fix and ensures a fresh, quality solution.

You judge quality solutions on whether they advance innovation goals and meet criteria. The importance of knowing the goals and criteria seems self-evident, yet many people do not consider these when solving R&D innovation problems. They accept the first adequate solution, the **quick fix**. The quick fix doesn't ever produce the highest quality solution no matter how good it may appear at first. Only careful, deliberative use of creative thinking techniques ensures the highest quality solution.

Everyone wants and needs to achieve high quality innovation solutions to R&D problems. innovation success depends on it. Yet relatively few R&D people have access to the new creative thinking techniques that lead to the highest-level solutions. These techniques focus on the real issues within an R&D innovation problem and on its quality solution, instead of on the preconceived ideas and notions with which we all (unawarely) limit ourselves.

---

This book describes these techniques step by step. Although the author recognizes the role of unstructured creativity in the problem-solving process, truly high caliber solutions come from proper use of modern techniques, some developed by the author, that spark different perspectives, fresh ideas, and innovative approaches.

This exceptional book shows how to focus on the high quality innovation solution, not on creativity. It focuses on techniques, not on mindless, enthusiastic fluff. It focuses on what works, not wishful theory. It focuses on innovation success.

Some of this unique book's other distinguishing features include:

• It presents many techniques simply and clearly within a logical and interesting sequence. Written in an informal, authoritative style, it motivates the R&D reader to apply this workable approach.

• It starts with the author's successful methods that create quality solutions in problem solving "Creativity & Innovation Meetings," R&D meetings that attack and solve impactful problems by pulling people together from all over the organization. To help produce quality solutions, subsequent chapters reveal techniques that create a creative climate, shift paradigms, define problems, generate ideas, and combine them into quality solutions for innovations in your R&D team, or while working alone.

• It presents true R&D stories and case histories of how people used the author's approaches to solve important innovation problems in Fortune-500 companies.

• It presents a simple eight-step problem-solving sequence and shows how to carry out the three key creative steps, describing techniques in clear step-by-step language.

• It has a separate section with techniques for R&D people who **work alone**.

• It has a separate section with techniques for R&D **leaders** who want to adjust their leadership styles to motivate R&D people to achieve quality innovation solutions.

This uncommon book sets a new standard for books on creative thinking and innovation in R&D. It takes the novice by the hand, and helps the expert hone existing skills. Its step-by-step approach takes the guesswork out of achieving quality solutions and reveals many insider secrets used by consultants.

## HOW TO USE THIS R&D BOOK:
## YOUR PARTICIPATION IS WANTED

You can read this book in many ways.

You can turn it into a quick read and not do any of the activities. People who have done this have told me it helps creative thinking by shifting paradigms and solving problems creatively.

Or you can read this book slowly, participating in the activities. People have told me this thoughtful approach is also extremely helpful to creative thinking.

Or you can skip around. Or use it as a reference. The choice is yours.

The way I intended was for you to learn by doing, to enjoy the experience of this book, and to capture the essence of creative thinking and solving innovation problems creatively by participation. Here's what else you will find.

You will learn to use three types of techniques to help solve R&D innovation problems creatively. This includes techniques to shift R&D paradigms and produce unexpected new ideas; techniques to change the R&D climate so new ideas flourish; and techniques to stop pigeonholing R&D people, including yourself, so you will stop stifling creative thinking.

You will discover advanced techniques to carry out the three key creative steps to shift paradigms and solve R&D innovation problems more creatively.

You will learn how to conduct targeted R&D problem-solving Creativity & Innovation Meetings.

You will learn special creative thinking techniques for Special R&D Creative Teams.

You will discover techniques to motivate R&D people to think creatively at work.

You will illuminate your habits that spoil creative thinking, and the techniques to deal with them.

You will find techniques to help the successful submission of new ideas and proposals.

If you are a R&D leader, you will learn techniques to adjust your leadership style to help creative thinking at work.

This book can work for you. The results will ultimately depend on your commitment to alter habits and apply new techniques to shift paradigms and solve problems more creatively in R&D at work.

---

# • PART 1 •

## ADVANCED CREATIVE THINKING TECHNIQUES SHIFT R&D PARADIGMS

-------------------------------------------------

### • CHAPTER 1 •
### PATHWAY TO INNOVATION:
### CREATIVITY & INNOVATION MEETINGS ATTACK MAJOR PROBLEMS
### FOR THE SUCCESS OF YOUR R&D ORGANIZATION

---

Whether you use advanced creative thinking techniques in team meetings, in permanent creativity teams, or when you work alone, you will achieve higher quality solutions to important innovation problems in R&D. These techniques work even better in a **Creativity & Innovation Meeting**.

I consider Creativity & Innovation Meetings the most effective way to **solve** important R&D innovation problems at work. Creative thinking facilitates the recombining of old ideas and elements in your mind into unexpected new and useful solutions. Advanced creative thinking techniques hasten and expand this process; this helps you combine more diverse bits from your environment and in your mind into new outstanding and useful ideas.

Creativity & Innovation Meetings attack R&D innovation problems so important and pervasive that management includes participants from many parts of the organization, and sometimes from outside the organization. This blending of an organization's people with modern techniques creates excellent innovation results. High quality solutions appear after an intentional effort to generate and select from a wide range of novel perspectives and ideas that create new possibilities.

Participants form teams of six to seven people (creative thinking teams), define the problem innovatively, generate hundreds of ideas, produce exciting proposals (one by each participant), and create high quality solutions (one by each team). The usefulness and creativeness of the quality solutions dazzle and fascinate.

I also consider these meetings the most effective way to **teach** creative thinking techniques. R&D people learn them because they want to achieve quality solutions to impactful innovation problems of the organization for their own success. They produce

---

ideas no one thought of before the meeting, not peddle notions already familiar to the people present.

---

A SUCCESS STORY: After attending one of my creative thinking workshops, a manager in a Fortune-500 company asked me to help him creatively solve problems in a new area assigned to him. He had approached the assignment in time worn ways. He wanted to change that and generate fresh ideas and new paradigms. We planned a three day Creativity & Innovation Meeting with five people from R&D, five from marketing, and five from manufacturing.

In the meeting, I mixed the participants into three teams and showed them how to apply advanced techniques to problems presented by the manager. By the time this meeting ended, the three teams generated over eight hundred ideas and wrote fifteen one-page proposals to solve his problems, one proposal from each person.

He later wrote me: "One of the approaches identified at the meeting has been picked up and should be commercialized shortly. We have had a very positive response from a major retail chain."

---

### Specific Problems

Tackle almost any innovation problem in a Creativity & Innovation Meeting. I have led meetings seeking quality solutions for diverse, mostly R&D, problems, including how to...

• identify new products

• solve mutual problems with special customers

• increase chemical yield during a complex manufacturing process

• reduce waste

• design a profitable and environmentally safe chemical plant

• apply world-class manufacturing principles to a specific product

• improve quality

• lower costs and increase effectiveness of environmental cleanup

• develop a new technology for manufacturing

• handle manufacturing waste

These meetings resulted in very successful outcomes.

# SOME SPECIFIC R&D CREATIVITY & INNOVATION MEETINGS

## Increasing Chemical Yield In A Six Day R&D Creativity & Innovation Meeting

I led a six-day Creativity & Innovation Meeting for a Fortune-500 company that wanted to identify and develop new approaches and new ideas to increase the yield of a chemical manufacturing process.

We conducted the meeting in a hotel near Orlando, FL for 25 managers, supervisors, scientists, and engineers in R&D and manufacturing. The meeting started at 6:00 p.m. on Sunday and continued until 1:00 PM the following Friday. It lasted from 6 to 10 p.m. on the first day, and continued each morning from 8:00 AM until noon, during which time we defined the problems and generated new ideas using advanced techniques.

Each day at about 12:30 PM, busses whisked people off to spend the afternoon and evening at Epcot Center, Sea World, or the Kennedy Space Center. I had shown people how to make and use metaphors to spark creative thought, and each day I asked people to find trigger-ideas and metaphors on their outings to help solve the problem.

The results gratified.

First: The participant's evaluations of the event were very positive.

Second: They generated high quality ideas and perspectives to solve the R&D problem.

Third: Almost everyone said that they would use the advanced creative thinking techniques back on the job. Thus, the company benefited by the diffusion of many new advanced creative thinking techniques into their work force.

Fourth: Everyone appeared eager to work and play each day, a welcome sign that all was well.

Fifth: I was asked to visit other R&D parts of the company and lead a three or four day local version of this Creativity & Innovation Meeting.

I suspect this meeting was successful partly because the design and presentation was based on design principles I describe in chapter 18, particularly the long length of this meeting and extensive use of daily adventures and incubation time.

Thus, this meeting benefited one Fortune-500 company by uncovering new approaches to solve a long-term chronic R&D problem, by diffusing new techniques within their work force, and by discovering how to enhance creative thinking at work.

**Improving Quality At Work In A Three Day R&D Creativity & Innovation Meeting**

A Fortune-500 company asked me to lead a three-day problem-solving Creativity & Innovation Meeting on how to improve 'quality' at work. This event took place in a beach resort hotel during off-season. The participants included 35 managers and supervisors in manufacturing and R&D from a large plant of this company.

I customized the design of this event to coincide with their goals for the meeting making sure to preserve its potential for success. Their meeting goals included:

• Each person will develop new perspectives and new ideas to help solve their own work related problems on 'quality.'

• Each person will develop a personal proposal to improve his or her own 'quality' at work.

• Each person will use and teach advanced techniques at work and solve problems more creatively (shift paradigms and define problems creatively, generate ideas super abundantly, and combine ideas into trigger-proposals and workable solutions).

The meeting consisted of six sessions.

• First session: We formed creative thinking teams, carried out advanced techniques to create a creative atmosphere, and started team building.

• Second session: Selected participants gave short talks to teach everyone the elements of 'quality' at work.

• Third session: We carried out advanced techniques to define the problems these people had about 'quality.'

• Fourth session: We carried out advanced techniques to generate ideas abundantly to help solve the 'quality' problems of this plant site.

• Fifth session: Each team helped solve work related 'quality' problems of each of its members using advanced creative thinking techniques learned earlier during this meeting.

• Sixth session: Each person wrote a one-page proposal to solve their own problems about 'quality' and received comments from their team to improve and upgrade the proposal. Each person then developed and shared their personal action plans to improve the 'quality' of their work output.

Excellent outcomes appeared. Each person left with personal action plans to improve quality in their work and to help other people improve quality too. And all learned creative problem solving techniques to apply at work. Outstanding.

**Improving Environmental Cleanup In A Four Day R&D Creativity & Innovation Meeting**

An Environmental Manager of a Fortune-500 company asked me to lead a four-day Creativity & Innovation Meeting to solve problems associated with environmental clean up at a particular plant. He held the meeting in Washington, DC.

Their goals and purposes:

- to generate novel and unexpected approaches to solve the plant's ground water and associated subsoil problems

- to improve current approaches

- to cut costs

- to apply the solutions to other plant sites

- to spawn new business opportunities

- to develop a broader thinking perspective and a more proactive view on how to deal with government policies and regulations

- to enable each participant to learn advanced creative thinking tools to enhance everyday thinking skills

- to boost synergies between the environmental problem solving capabilities within their company

- to develop acceptable remediation programs that met their environmental concerns and commitment to society without adversely affecting the financial health of their businesses

The agenda for the 4 days included:

• Session 1: We went through introductions, reviewed the goals and agenda, created a creative atmosphere, formed creative thinking teams and started team building, wrote and applied metaphors, poems, and trigger-ideas to environmental clean up problems.

• Session 2: We finished the work started in the previous session.

• Session 3: The Environmental Manager gave a short presentation on the clean up problems: our current ideas; what we do now; and what we want to do. We then defined environmental clean up problems creatively.

• Session 4: To expand their creative thinking processes, I asked everyone to visit the Smithsonian Air And Space Museum (and the Museum of Natural History or an Art Museum) and write metaphors and poems about the environmental cleanup problem. The evening, free time, for incubation.

• Session 5: We generated ideas to solve environmental clean up problems using advanced creative thinking techniques.

• Session 6: I asked everyone to write metaphors and poems in the Smithsonian 1876 Exposition (and in the Museum of Natural History or an Art Museum). The evening: Free time again.

• Session 7: We generated trigger-proposals, identified criteria to select ideas, and developed quality solutions.

• Session 8: We upgraded the environmental clean up proposals and developed specific action plans. The Environment Manager made serious commitments to support the proposals.

An outstanding outcome. They developed many new and unexpected ideas. The participants wrote very positive evaluations of the meeting.

**Developing Unconventional Manufacturing Technologies In A Three Day R&D Creativity & Innovation Meeting**

A Senior R&D Engineer of a Fortune-500 company wanted to develop new, unconventional technologies to make products that would leap frog the company over competitors for a profitable sustainable business. He asked me to design and lead a three-day Creativity & Innovation Meeting. He wanted an outcome that:

- upgraded their current approach and lowered costs

- identified seemingly unrelated technologies that would work together

- used advanced creative thinking principles

- shared group knowledge and experiences

- enhanced synergies between the problem solving capabilities within the company

- led to the identification of specific concept proposals and a plan of attack

- enabled each person to learn some creative thinking tools to add to everyday thinking
  skills

I asked each person to read any creative thinking book ahead of time, and to bring their ideas concerning the new technology with them along with a willingness to **replace** these pet preconceptions with fresh ideas.

Twenty-five R&D engineers and scientists attended this meeting in a hotel in Virginia Beach, VA. There were seven sessions, one an afternoon of free incubation time.

Session 1 consisted of the usual introductions and a review of the goals and agenda. We used techniques to create a creative atmosphere, started team building, and learned some techniques that spark creative ideas in teams.

Session 2 started with a short presentation of the problem by the Senior Research Engineer who discussed current ideas, what was going on now, and what he wanted to accomplish. After that, individuals, and then the teams, defined the problem creatively.

In session 3, the teams generated ideas abundantly, after which each individual generated ideas sitting quietly alone.

Session 4 consisted of an afternoon of free incubation time plus some individual fun work to enhance creative thinking.

In session 5, each person wrote a one-page trigger-proposal for a new technology. Each team identified the criteria to select ideas and then suggested ways to improve the trigger-proposals of its members. Each person then wrote a one-page workable proposal and gave it to the Senior Research Engineer in charge. Each team then combined the proposals of its members and developed a blockbuster proposal to solve the technological problem.

During session 6, each creative thinking team presented its blockbuster proposal to other teams so they could improve and upgrade the solution. They gave all ideas in writing to the Senior Research Engineer.

Session 7 consisted of developing final action plans and making commitments to implement them.

The meeting was a very successful. The participants generated over 975 ideas. Each person developed a one-page workable proposal for a new technology. Each of the three creative thinking teams generated a blockbuster proposal by combining and developing the trigger-proposals of its members.

Later, the Senior Research Engineer told me that 15 to 20 patentable ideas had emerged during this creativity meeting, and he asked me to lead another meeting to develop these new technologies.

**Improving Waste Handling In A Two Day R&D Creativity & Innovation Meeting**

Most meetings I lead are four days long. This length helps the incubation processes in the mind so unexpected connections are made and blockbuster ideas eventually appear. Three-day Creativity & Innovation Meeting also work, though not as well.

I therefore hesitated when I agreed to design and lead a two-day meeting for a manufacturing plant of a Fortune-500 company in Goldsboro, NC, on how to deal with manufacturing waste. Besides the District Manager, the meeting contained 15 managers from the plant.

They wanted to:
- improve their current methods and identify unexpected approaches
- introduce new ways to use creative thinking principles in management
- share knowledge and enhance synergies between different groups in the company
- build team participation
- enable each person to learn some creative tools to enhance everyday thinking skills at work

I designed this two-day meeting to counter the absence of the multiple incubation periods found in three- and four-day meetings. In addition, I designed the **front and back end** of the meeting carefully. The front end focuses on creative thinking while the back end focuses on the logical, evaluative thinking required to choose and develop quality solutions.

In a four-day meeting, I plan 2.5 days of creative thinking followed by 1.5 days of logical thinking. Usually, in a three-day creative thinking meeting, I plan two days of creative thinking, followed by one day of logical thinking. In this two day meeting, I planned one day each of creative and logical thinking. Too little creative thinking, I thought. I maximized incubation thinking overnight.

During the first session, we jumped into creative thinking, the **front end** of the problem-solving process. We went through introductions; discussed the goals and agenda; started team building; created a creative atmosphere; and used trigger-ideas in linear and nonlinear creative thinking. Then the District Manager introduced the problem.

During the second session, the teams defined the problem and generated ideas to solve it. Individuals then generated trigger-proposals to solve the problem. I asked the participants to work overnight and finish the trigger-proposals, thus stimulating their incubation thinking.

During the third session (the next morning), I started logical thinking (the **back end** of the problem-solving process). The teams identified the criteria to use to select ideas and proposals. Individuals then shared trigger-proposals to solve the manufacturing waste problem and their team upgraded it. Each team then developed a blockbuster proposal to solve the manufacturing waste problem that fit the criteria.

During the fourth session, each team presented their blockbuster proposals to solve the manufacturing waste problem for comment and improvement from the other teams.

The outcome of this two-day creative thinking meeting fascinated. Hundreds of ideas appeared, Each person handed in a one-page proposal to the District Manager, so he now had 15 proposals to apply to the manufacturing waste problems. Finally, each team produced a blockbuster proposal, so he had three different proposals for further consideration by his management team. One proposal completely shifted the paradigm, one proposal straddled the paradigm, and one proposal stayed creative within the current paradigm. Several people suggested that the District Manager could combine (and greatly improve) the three proposals.

Still, given a choice, I prefer a four-day meeting to a two day event. The longer incubation time for new ideas and the higher creative thinking level produces more quality ideas, more new paradigms by each group, and the final solutions achieve higher quality.

**Identifying New Products In a Three Day Creativity & Innovation Meeting**

I led a three-day Creativity & Innovation Meeting for 42 people from a moderate sized corporation in Ann Arbor, MI, to **identify new products**. The marketing vice president arranged the meeting. The Chairman of the corporation, its president, the vice-presidents and managers of sales, marketing, manufacturing, engineering, and finance, and the directors of personnel and quality attended. The rest of the 42 people present included key professionals in sales, marketing, customer relations, industrial design, manufacturing, engineering, and finance, a wide range of people chosen specifically to tackle the problem.

Before the meeting, we put the 42 people into six creative thinking teams of seven people each and asked each person to read different sections of a creative thinking book ahead of time.

The company wanted to...

- generate novel and unexpected ideas for new products

- identify new product concepts for further development

- learn creative thinking principles to create new approaches to problem solving

- enhance synergy and share knowledge within the corporation
- build team participation

   • The first session started with a discussion of the meeting goals by the chairman of the corporation, followed by introductions, beginning team building, creating a creative atmosphere, learning to use trigger-ideas to spark better ideas, and generating ideas for new products using simple techniques.

   • In the second session, people used advanced techniques to generate ideas about new products.

   • In the third session, people used special advanced techniques to generate about 750 ideas for new products and displayed them in the meeting room.

   • During the fourth session, each person looked over the ideas and selected and combined many of them into a one-page written trigger-proposal for a new product.

   • During the fifth session, each participant shared this trigger-proposal with his or her team for improvement. They revised the one-page proposals and gave them to top management for later consideration. Each team then developed its own blockbuster idea for a new product.

   • During the last session, each team presented its blockbuster solution on large flip chart paper to the other participants and received ideas for improvement from everyone, an exciting, constructive time.

   Very positive outcomes came forth.

- People wrote very positive evaluations.
- Top management received a large number of ideas for new products, among them real gems.
- Management received 42 one-page written proposals for new products, one from each participant, many of them hot and unexpected ideas.
- Each team produced one blockbuster idea for a new product.
- The corporation executives, all of whom attended, thought the time and money well spent.
- And the Chairman later wrote me "to declare the creative thinking sessions a roaring success. Besides the excellent ideas generated, and there are a handful of outstanding ones, it served the purpose of bringing our management team closer together.... Thanks for an outstanding job."

**An Engineering Firm Invites a Client To A 3 Day R&D Creativity & Innovation Meeting**

A consulting engineering firm and its client, a chemical company, collaborated in a three-day Creativity & Innovation Meeting that I led in Denver, CO., applying creative thinking techniques to the design of **environmentally safe** chemical plants.

We initially held a one-day planning meeting in a plant of the chemical company. We agreed on the goals of the meeting:

- to generate ideas and written proposals to make a chemical plant environmentally safe

- to upgrade their current approach

- to identify new approaches

- to broaden their thinking about solving problems creatively

- to share group knowledge

- to build synergies and team participation

- to teach each person some creative tools to enhance everyday thinking skills

Before the meeting, each participant read a creative thinking book of their choice, and brought with them a willingness to replace pet preconceptions with new ideas.

The meeting started at noon with an informal lunch, during which representatives of the two companies explained the problem and the goals of the meeting. I then took over.

The first session consisted of introductions, reviewing the goals and the agenda, creating a creative atmosphere, starting team building, learning the use of trigger-ideas and metaphors, and using some advanced techniques that spark creative ideas.

During session 2, the planning team explained the purpose of the meeting and the participants defined the problems creatively.

In session 3, the teams generated ideas using many techniques, after which each individual generated additional ideas sitting quietly alone.

Session 4 consisted of an afternoon of free incubation time writing metaphors and trigger-ideas to spark creative ideas.

In session 5, each person wrote an innovative one-page trigger-proposal for new approaches.

During session 6, the planning team presented the criteria to select ideas. Each team suggested ways to improve the trigger-proposals of its members, who then wrote a one-page workable proposal that they handed in. Each team then combined the ideas of its members and developed a blockbuster proposal.

During session 7, each team presented its proposal so the other teams could help improve them. These proposals were also handed in. Finally, the people from the engineering company planned how to spread creative thinking throughout their company.

The written evaluations, so very positive. Everyone had learned to apply effective techniques to achieve quality solutions to problems at work. They generated about 950 ideas. In addition, they generated 30 one-page written proposals, one from each participant. And each of the five teams generated a blockbuster proposal by combining and developing the trigger-proposals of its members. The people present thought the time (and money) well spent.

The blending of people from these two companies in a focused endeavor to create powerful proposals had a very serious purpose, and produced excellent, practical outcomes.

## • **Conclusion**.

This chapter describes the advanced creative thinking techniques needed for R&D Creativity & Innovation Meetings. These meetings have a huge payoff producing high quality solutions to important innovation problems.

I described six specific R&D Creativity & Innovation Meetings above. As you continue reading this book, you will encounter the techniques you will need:

- paradigm shifts (Part 1),

- creating a creative climate in your mind and R&D workplace (Part 2),

- using techniques within systematic problem-solving steps that lead to fresh, high-quality solutions for R&D success (Part 3)

- creative thinking in R&D meetings (Part 4).

Still, achieving quality solutions for innovations at work means more than meetings and teams. You will need:

- techniques to use when you work **alone** (Part 5).

- leadership approaches that foster the creative thinking that leads to quality solutions from R&D people (Part 6).

- habits and techniques that affect creative thinking and discover ways to help, not hinder creative innovations.

These approaches will enrich your R&D innovations.

---

**· CHAPTER 2 ·**

**A PARADIGM SHIFT: THE SOUND OF ONE HAND CLAPPING**

If you always think what you have always thought,

Then you will always do what you have always done,

And you will always get what you have always gotten.

---

A paradigm is a belief structure within which you think and act. The paradigms within which you operate affect your creativity. Usually they box you in and produce tunnel vision. A paradigm shift is a change in your belief structure that changes your perspective and allows you to see things differently.

An old Zen riddle asks, "How do you get through a gateless gate?" Do you mean how do you accomplish the impossible (a paradigm shift)? Well, you start by defining the problem creatively and shifting paradigms, by listing many ideas, and by combining them innovatively into creative trigger-ideas and workable solutions. Committed action plans and the real work follow.

---

**A RIDDLE: HOW DO YOU GET THROUGH A GATE LESS GATE?**

You mean, how to do the impossible at work?

(Note the paradigm shift)

Well, you start by defining the problem creatively, shifting paradigms, generating ideas abundantly, and combining them innovatively into creative solutions. After that comes committed action plans and the real work.

---

At the beginning of my creative thinking workshop, I often ask "What is the sound of one hand clapping?" Most people give no answer. Some answer the "sound is silence," a breakthrough in realizing that the answer does not have to be clever.

When I ask children this question they look puzzled. Sometimes they wave one hand through the air listening for the answer. Actually, the waving of one hand is one answer to the question. This is hard for some people to understand.

In some Zen training, this riddle of the "sound of one hand clapping" is given to novices starting to master Zen. The novice meditates on the meaning of the riddle, and makes daily visits to the Zen master for about three years to absorb the riddle's teachings.

According to Yoel Hoffman in "The Sound of One Hand: 281 Koans with Answers," the acceptable answer is for the novice to face the Zen master, take a correct posture, and silently extend one hand forward. This answer embodies much of Zen philosophy. It is immediate, nonverbal, spontaneous, and intuitive, and so is creative thinking.

The Zen answer has many nuances that we need not pursue. Suffice to say that three years of meditating on the "sound of one hand clapping" produces a paradigm shift in the novice's view of reality.

## SHIFT PARADIGMS TO ACHIEVE QUALITY SOLUTIONS

My purpose in using the riddle about the "sound of one hand clapping" is to produce a quick paradigm shift to help creative thinking. Paradigm shifts help creativity.

I want to jolt people into realizing that the way we perceive a problem limits our thinking. Almost immediately, some participants discover some creativity-spoiling habits that block their creative thinking. This discovery prompts a change in perception of creativity, and how to enhance it. These people discover that ideas can be expressed nonverbally, as well as in writing. They see the value in being spontaneous and intuitive, as well as rational. They see the need to be immediate, as well as reflective. They see that creative thinking can be helped by changes in perception.

All this is triggered by a Zen riddle. This book is also like a Zen riddle in that it is intended to change your perceptions, produce paradigm shifts, and enhance your creative thinking with advanced creativity techniques.

---

A TRUE STORY: This chapter is the result of trigger-ideas and paradigm shifts sparked by reading about a dozen books on Zen. These trigger-ideas and paradigm shifts evolved into ideas that I incorporated into the design and activities in my creative thinking workshops and throughout this book. I did not interrupt the text to point out Zen applications because I did not want to distract from the flow of ideas presented and because the Zen trigger-ideas usually evolved into outcomes unrelated to its Zen origins.

---

**WHAT IS THE SOUND OF ONE HAND CLAPPING?**

Well, it's like the sound of one brain thinking creatively (note this paradigm shift).

Oh, you mean it is a silent explosion in the universe.

---

## • CHAPTER 3 •

## IF YOU ALWAYS DO WHAT YOU HAVE ALWAYS DONE, YOU WILL ALWAYS GET WHAT YOU HAVE ALWAYS GOTTEN

---

Creative thinking fuels your ability to generate an unexpected "new and useful" idea. The more unexpected the idea, the more creative we perceive it. Now it can work even better. An important paradigm shift occurred.

### THE QUIET REVOLUTION IN CREATIVE THINKING

For the past 75 years or so, a quiet revolution has taken place. No longer do we have to wait for someone's brain to slowly churn out a new and useful idea. Now we have hundreds of creative thinking techniques to help generate ideas to achieve high-quality solutions to problems.

"Hold on," you say. "Didn't the old techniques work for thousands of years to produce new and useful ideas? Didn't we construct our entire civilization using those old ways? Why a new way? If it ain't broke, why fix it?"

Of course the old ways still work. And we did construct our civilization waiting for new and useful ideas to slowly appear.

Nonetheless, the new ways heat up the process so we generate more new and useful ideas in a shorter time. And because we have more ideas to choose from, we turn out higher quality solutions. Quantity of ideas often leads to higher quality solutions, and avoids the quick fix.

R&D teams using brainstorming to generate ideas complain that they have trouble sorting and selecting so many ideas. How sweet to move from the old bottleneck of not having enough ideas to the problem of evaluating the myriad of ideas a simple brainstorming session produces.

And Osborn invented brainstorming over 80 years ago, a time when we used the Model-A Ford and the DC-3 airplane. Not a modern procedure at all. Indeed, using only brainstorming today mimics driving around in a Model-A Ford or a DC-3 airplane, ignoring computers, antibiotics, TV, the Internet, the I-pad, and the thousands of modern inventions from which we benefit.

---

Most people do have a choice when solving problems. Unfortunately, while seeking the highest quality solution, they use old fashioned methods with a lower probability of success. They ignore modern creative thinking techniques.

Put another way, the old fashioned way of waiting for ideas mimics evaporating water by letting it sit in a large dish. Modern techniques add heat to the process, and some state-of-the-art techniques act like a blast furnace. Really, brainstorming is old fashioned today.

So move into the 'Age Of Modern Creative Thinking Techniques.' Start using the many dozens of creative thinking techniques now available to help you achieve quality solutions. I describe many throughout this book.

## THIS BOOK WILL HELP

This book focuses on what works, not theory. I presented workshops for over 15 years. I learned that while theory help us understand, techniques that work at work help us even more.

I once had a great deal of skepticism about people's ability to learn creative thinking skills. My scientific colleagues did not believe it could occur. Also, I showed no improvement in my ability during the previous three decades. Neither did my colleagues and students who worked with me during those decades. Our skepticism seemed well based in data.

In retrospect, our lack of improvement seems reasonable: none of us knew how to improve our creative thinking skills and we didn't think we could do it, so we didn't do it -- a circular trap,

What changed my opinion? In 1983, while a Visiting Fellow at the Center For Creative Leadership, I developed my workshop on creative thinking skills. Participants reacted very positively. These included executives, managers, supervisors, department heads, directors, professionals, and others from many fine corporations. Their responses persuaded me to focus my workshop on advanced creative thinking techniques to achieve high quality solutions at work, a theme that permeates this book.

These techniques helped me and the thousands of people who attended my workshops. I emphasized three types of techniques to help solve problems creatively.

First, use creative thinking techniques to shift R&D paradigms and produce unexpected, fresh ideas and solutions.

Second, use techniques to change the R&D climate so new ideas flourish.

Third, use techniques to stop pigeonholing people and stifling their creative thinking.

I describe all three types of techniques in this book. You will also find the following:

• Techniques to carry out the three key creative steps in R&D that shift paradigms and achieve creative solutions.

• Techniques to conduct R&D Creativity & Innovation Meeting.

• Techniques for permanent R&D creative thinking teams formed from employees in your organization.

• Techniques to motivate R&D people to create creative outcomes at work.

• Habits that spoil creative thinking in R&D, and the techniques to avoid them.

• Techniques to help 'sell' new R&D ideas and proposals.

• Techniques to adjust your leadership style to help the creative thinking of other people.

This book can work for you. The results ultimately depend on your commitment to alter habits and apply new techniques to R&D problems.

## TO ACHIEVE SUCCESS, REDUCE RESISTANCE TO NEW TECHNIQUES

Today, many companies find that creative thinking contributes to a winning competitive edge. R&D teams and people who work alone use techniques to solve problems more creatively. Turn your complacent R&D team into a dynamic creative unit using advanced techniques.

Most people drop resistance once they accept some new ideas:

First, creative thinking in R&D involves an ordinary, daily activity.

Second, R&D people can learn creative thinking techniques. Using these techniques does not turn anyone into an Einstein, but they do help people find better solutions to problems.

Third, people can change the habits that spoil creative thinking and doom the creative climate. Finally, using techniques helps achieve quality solutions and leads to success.

Skeptical R&D people need to discover that many different ways of perceiving a work problem exists, that diverse solutions exist, and that using techniques works, a key change in attitude.

So, if you want to do better, if you want to move toward excellence, and if you want more profitable R&D products and services, start using advanced creative thinking techniques to improve your winning competitive edge.

This book will show you how to do it more effectively.

## CREATIVE THINKING TECHNIQUES DO HELP

Many R&D people underestimate the power that creative thinking techniques provide because they persist in retaining unrealistic ideas, especially if they believe creative thinking comes from an exceptional, inherited gift. You either have it or you don't.

Not so. Most of us have this ability and use it everyday. We don't recognize it as creative thinking or even see it as special. We call it tinkering, ingenuity, intuition, trial-and-error, imagination, making suggestions, inventing: anything but creative thinking. We think creative thinking an exceptional gift inherited by other people.

Not true. Most R&D people think creatively most of the time; it depends on what you spend your time creating that makes the difference. Best of all, creative thinking techniques helps solve problems more effectively in all areas of your organization.

Many levels of creative thinking exist, from low daily levels to hot, unexpected levels. Increase the probability you operate at a higher level by using advanced techniques to create a creative atmosphere in your mind and in R&D without pigeonholing yourself or other people.

## A MYTH ABOUT CREATIVE THINKING IN THE INNOVATION PROCESS

Some R&D people think that the innovation process only requires creative thinking during the generation of the big-bang idea. After that, its hard, dull work. Purely a myth. Creative thinking solves daily problems throughout the process of innovation. Usually we disguise what happens by calling it something else like tinkering or fooling around.

On-the-job creative thinking spurs the daily, ongoing process of transforming old ideas into new ideas to solve daily problems throughout the R&D innovation process.

---

A TRUE STORY: One Fortune-500 organization invited three people who worked for their best customer to attend one of my workshops on advanced creativity techniques. As they put it: "We wanted to solve our mutual problems creatively, but couldn't because they didn't know about creative thinking techniques."

Last I heard they were going to invite their main suppliers as well, so they could shift paradigms and solve mutual problems together up and down the line.

---

---

# • CHAPTER 4 •
## CREATIVE THINKING: INHERITED OR LEARNED

If it isn't inherited, where do creative thinking skills come from?

---

I trained as a geneticist, and "Is it inherited?" always intrigues me. After all, if we inherit creative thinking like I.Q., then we can do little to help a low creative person.

Some creative thinking skills needed at work include the following abilities. Do we inherit them?

• The ability to keep an open mind, switch tracks, see new perspectives, shift paradigms, and generate different mind sets.

• The ability to associate remote stimuli in the environment with elements in the mind and combine them into new and unusual ideas.

• The ability to generate many ideas.

• The ability to use many different problem-solving approaches.

• The ability to generate a variety of really different ideas.

• The ability to develop ideas.

• The ability to generate infrequent and uncommon ideas.

• The ability to hang in there when going against consensus and being persistent in the face of criticism.

To me, few of these skills seem inheritable and most seem learnable with appropriate training. Indeed, research comparing the scores of identical and fraternal twins on tests of these skills show that creative thinking does not have a large genetic contribution. You can find the evidence in the articles listed at the end of this chapter.

Good news. If we inherited our creative thinking skills, this book would end now with condolences: sorry, you will always have what you have now, and training in creative thinking will not help. Instead, most people with reasonable mental ability become more creative after they learn creative thinking techniques.

Thus, escape from a major myth that creative thinking has a large inherited component. This myth leads to a self-fulfilling prophecy: nothing can harm a high-creative person and nothing can help a low-creative person. Not so on both accounts.

---

First, as shown by the research on identical and fraternal twins, we do not inherit most creative thinking skills.

Second, we can learn creative thinking techniques.

Third, the workplace environment and other external factors can help or harm creativity.

Finally, success as a creative person also depends on your other talents and abilities (inherited or otherwise), your motivation, and your interpersonal skills. The quality of your relationships in R&D, especially with your boss, affects your creative output and commitment to innovation.

---

A HABIT THAT SPOILS R&D CREATIVE THINKING: Thinking you will always **lack** enough creative skills. This negative put-down of yourself spoils your attempts to achieve creative outcomes. Instead, use advanced creative thinking techniques.

---

## RESEARCH SHOWS WE DO NOT INHERIT CREATIVE THINKING SKILLS

Richmond, B.G. (1966) Creative thinking in monozygotic and dizygotic twins. Research Report from EDRS (ERIC Document Reproduction Services of the U.S. Dept. of Health, Educ, and Welfare, National Institutes of Education).

Torrance, E.P. (1976) The creative child in the family; each one is "special." The Creative Child And Adult Quarterly, Winter, 195-199.

Pezzullo, T.R. (1970) The genetic components of verbal divergent thinking and short-term memory. Ph.D. Thesis, Boston Coll., 92 pages.

Pezzullo, T.R., E.E. Thorsen, and G.F. Madons (1972) The Heritability of Jensen's Level I and II and Divergent Thinking. Amer. Educ. Res. Jour. 9: 539-546.

Reznikoff, M., G. Domino, C. Bridges, and M. Hoffman (1973) Creative abilities in identical and fraternal twins. Behavior Genetics 3: 365-377.

Domino, G., J. Walsh, M. Reznikoff, and M. (1976) A factor analysis of creative thinking in fraternal and identical twins. Journal Genetic Psych 94: 211-221.

---

• CHAPTER 5 •

## TECHNIQUES TO STIMULATE YOUR CREATIVE THINKING

You Want To Be More Creative Than Your R&D Competition, Don't You?

---

## HOW DO WE DO CREATIVE THINKING?

How do we humans carry out this wondrous activity called creative thinking? One intriguing notion suggests that chance and external triggers bring together diverse elements in your environment and in your mind into one thought, and this connection triggers creative solutions. We call this process 'making remote associations,' suggesting the need for a prepared, active mind full of diverse elements.

As Pasteur pointed out: "Chance favors the prepared mind." One way to become more creative includes preparing your mind with new bits for creative thinking by attending trade fairs, meetings, reading, travel, talking to peers, customers, vendors, etc.

Thus, on-the-job R&D creative thinking consists of **combining** old information and old ideas into unexpected new and useful ideas. The more unexpected, the more creative. This thought runs contrary to the myth that creative thinking creates new ideas out of nothing. In other words, solving problems creatively involves a down-to-earth activity, not a mountain top phenomenon.

Recognizing creative thinking as an ordinary act that combines and transforms old information into new ideas allows us to accept creative thinking as a natural process. You don't need special inherited gifts to use advanced techniques to solve problems creatively.

## INCUBATION AND OTHER STAGES IN THE R&D CREATIVE PROCESS

Some stages of the R&D creative & innovation process include the following:

• **Preparation Stage**. Fact finding; laying the groundwork and learning the background; learning the creative process.

• **Concentration Stage**. Total absorption in the problem; trancing out.

• **Incubation Stage**. Taking time out; resting; seeking distractions; working on other things; vacationing; jogging; taking walks.

• **Illumination Stage**. The AH-HA insight forms and ideas pop out.

---

• **Implementation Stage**. Solving practical problems of implementation; getting other people involved. In other words, the hard work.

The preparation stage, during which you fill your mind with new elements to make remote associations later, can last many years: in school, on-the-job training, reading, taking courses and workshops, traveling, life experiences, etc. After all, you cannot be a creative chemist, engineer, or computer whiz unless you know chemistry, engineering, or computers. You learn your craft or profession first.

During the concentration stage, you focus on a particular problem and absorb yourself in it, making a place in your mind for a new idea to enter.

Frustration at not finding a solution leads to the incubation stage, during which you concentrate on other things while your mind takes a break and quietly makes remote associations.

Then, if you are fortunate, the illumination stage occurs, the paradigm shifts, the AH-HA insight forms, and a new idea emerges.

Then the implementation stage occurs, a stage that can last a short time or a lifetime, as the entire process cycles repeatedly to modify, implement, and develop the idea into a profitable innovation.

Thus, new R&D ideas do not appear spontaneously out of the blue. They require preparation, concentration, incubation, and the appropriate triggers to spark remote associations. When new ideas appear, they need special and deliberate nurturing or they disappear.

These notions trigger a number of issues at work.

• How much incubation time do you build into your schedule? Would your organization pay you to spend a day or two walking in the woods or sitting on a beach?

• If someone sits with his or her feet on the desk looking out the window for several hours, or even days, would people in your organization find this behavior acceptable?

---

A HABIT THAT SPOILS R&D CREATIVE THINKING: You do not allot enough **time** to the incubation stage of the creative process.

---

• Is "doing things" more highly valued than "thinking"? Is it okay for R&D people to spend time thinking, that is, seeming to do nothing?

• How much time do you allot to the preparation stage to get additional diverse elements into your mind?

• How do you obtain diverse elements for your mind? By travel, meetings, training, reading, conventions, trade fairs? By talking to customers, suppliers, competitors, people in other companies, in foreign lands, in other professions? Does your organization encourage and pay for this?

---

A HABIT THAT SPOILS R&D CREATIVE THINKING: You do not act to increase the **diverse elements** in your mind.

---

## BIZARRE TRIGGER-IDEAS SHIFT PARADIGMS AND SPARK BETTER IDEAS

Creative thinking yields many non-useful, even bizarre, ideas. All ideas, even bizarre ideas, can act as useful stepping stones to trigger better ideas, and to even shift paradigms.

For example, consider the statement: "Let's train bears to climb telephone poles in winter and shake off the ice that breaks the transmission wires." One person proposed this idea to prevent ice from breaking power lines in a mountainous region where winters are cold and rainy. Someone in the meeting had complained about being harassed by bears on one repair trip. This led one of the people, in a spirit of fun, to suggest training bears to climb the poles and shake the ice loose, clearly a bizarre idea.

A second person, again in jest, suggested putting pots of honey on the top of the poles so the bears would climb the poles and shake the ice off the wires, another bizarre idea.

A third person suggested, still in fun, using helicopters to place the pots of honey on the poles to attract the bears, also a bizarre idea.

Yet this led to a solution worth testing. The down draft from helicopters flying over the wires might knock the ice off. *

---

\* See "The Honey Pot: A Lesson in Creativity & Diversity" by Elaine Camper, April 2, 1993, for another telling of this tale at  http://www.insulators.info/articles/ppl.htm and also here: "Reengineering Tool Kit: 15 Tools and Technologies for Reengineering Your Organization" by Cheryl Currid, Prima Publishing, 1996, 320 pages.

---

People embellish this story with each retelling. Still, it offers wisdom: bizarre trigger-ideas spark useful solutions. Unless you encourage and help bizarre ideas to survive in R&D, you will hinder creative thinking and lose opportunities for innovation.

A HABIT THAT SPOILS R&D CREATIVE THINKING: We **squelch** bizarre trigger-ideas instead of using them to shift paradigms and spark better ideas.

## LINEAR AND NONLINEAR CREATIVE THINKING

Many ways to get new and useful ideas in R&D exist. One way uses **linear** creative thinking. We use this most because of its low risk. It flows like this...

A => B => C => D => E => New And Useful Idea.

You check each step carefully for truth and logic before moving to the next. You know your direction, how to get there, when you get there, and why you wanted to get there in the first place. Very precise, analytical, and certain.

Does it work? Of course. We base most of our rational thinking on this model. We constructed much of our civilization using this approach.

Another way to get new and useful ideas uses **nonlinear** thinking. Ideas leap about.

A => L => Z => R => E => X ... and eventually out of bizarre trigger-ideas and very remote associations, a paradigm shift occurs and some new and useful ideas emerge.

A very uncertain process. You do not know the direction, how you will get there, when you get there, or why you wanted to arrive there. Risky, unpredictable, and ambiguous, it often leads nowhere. But when a new and useful idea emerges, it likely represents a paradigm shift, unique & original.

Look at the nonlinear creative thinking path that turned the bears and the pots of honey into the helicopters.

• Repair man reports harassment by bears.

↓

• Train bears to climb poles and shake ice off wires.

↓

• Place pots of honey on top of poles to attract bears to climb.

↓

• Use helicopters to place pots of honey on top of poles to attract bears.

↓

• Use the down blast of helicopters to shake ice off the wires.

None of these bizarre ideas logically led to the other, yet it produced a testable outcome. Had the chairperson or anyone else at the meeting stopped the process or insisted on seriousness, the paradigm shift and the useful innovation would not have occurred.

---

A HABIT THAT SPOILS R&D CREATIVE THINKING: We don't deliberately **misperceive** the world to obtain a nonlinear viewpoint and a paradigm shift.

---

Many creative thinking techniques described in this book depend on you allowing trigger-ideas to flourish and eventually spark unique and sensible solutions to problems in R&D at work. Help innovation through bizarre trigger-ideas.

# • PART 2 •

## TECHNIQUES TO CREATE A CREATIVE ATMOSPHERE IN R&D

---------------------------------------------------------

### • CHAPTER 6 •

### HABITS THAT INTERFERE WITH THE
### CREATIVE ATMOSPHERE IN YOUR MIND

A TRUE STORY: I have a friend in R&D who uses his constant, gentle wit and says funny things. His humor helps when the discussion becomes overly serious. "Everyone thinks I'm humorous," he says. "Actually, I'm out of control." Indeed, he acts very spontaneously.

A HABIT THAT SPOILS R&D CREATIVE THINKING: We **inhibit** our spontaneity and **repress** our wit and humor.

When solving problems, instead of shifting the paradigm and striking out into new territory, we tend to turn to a former successful approach, Our mind focuses on the previously successful solution and we continue the old time worn ways of doing things even if counter productive. This dampens the creative atmosphere in your mind.

The antidote to this habit starts with becoming aware of this process and taking measures to reduce or eliminate it. That's one purpose of Part 2 of this book.

Let us start with some fun puzzles to discover how we spoil creative thinking in ourselves and other people, and the antidote to such spoiler-habits.

## IX

Each of us has far more creative thinking ability than we suspect. A bit of fun will show you how.

**Add one line to a "IX" and turn it into a six**. Spend about three to four minutes on this problem before moving on. No peeking at the answers yet, please.

ONE ANSWER: There are many solutions. The most common: add an "S" to the "IX" and produce a ... SIX

If you got this, congratulations.

If you did not get this answer, why not? Like many people in my creative thinking workshop, one or more of the following may have blocked you:

- You forgot that words can express numbers.
- You looked for a straight line. You forgot that lines also curve.
- You connected this in your mind with a match stick problem.
- You got stuck on Roman numerals.

Our thoughts act like they get trapped by collectors in our minds and cannot get out. And out of habit, we keep trying to find a solution within these collectors even though they do not work to solve the new problem.


## MIND FUNNELS

I call them mind funnels, collectors that capture R&D problems. Once you get stuck in a mind funnel, you find it hard to get out without deliberate creative thinking, that is, without using special creativity techniques. Every time a related new R&D problem arises, you return to the mind funnel that succeeded before. If you stuff a new problem into an old mind funnel that once worked, you generate the same time worn solution. In this sense, mind funnels act like paradigms.

Since a mind funnel gets bigger each time you use it to solve a problem, you eventually have little choice about how you perceive and deal with a new problem. That huge mind funnel captures your problem, and you exert your thinking efforts to push the problem through to an adequate solution, a quick fix, instead of seeking alternative mind funnels. You need to shift paradigms using techniques to get out of old mind funnels.

Thus, to convert "IX" to "SIX" you need to pass through at least two mind funnels: one that tells you "words can express numbers," and another that tells you "lines can curve." If you don't do that, you will not get to "SIX" from "IX."

---

A HABIT THAT SPOILS R&D CREATIVE THINKING: We allow our thoughts to get stuck in an old established **mind funnel**, and we stay stuck in **ruts** and old paradigms. We do not deliberately search out **alternative** mind funnels and shift paradigms using appropriate techniques.

---

You might have shifted the paradigm and gotten into other mind funnels and other answers, such as:

• Add a 6, and make 1X 6 (one times six). This equals 6.

• Cover the top half of the IX with a thick line, and turn it upside down so it looks like this: vi

• Move the vertical line in IX to the right and one slanting line of the X to the left to produce a distorted V|

• Fold the paper through the middle of the IX, and turn it over so all you see is VI

These last two solutions may disturb you because I did not add a line. Try to discover what mind funnel(s) grabbed your thoughts. Perhaps the following:

    • Fairness: I said add a line, and it seems unfair not to add one.

    • Making Unwarranted Assumptions: You probably assumed the added line must attach to the answer. Actually, I did not specify where or when you add the line, perhaps on the next page, or in the next edition of this book, or you could add it next week. If it disturbs you that much, please add a line just below the solution like this: <u>VI</u>

---

A HABIT THAT SPOILS R&D CREATIVE THINKING: Solutions to problems have to seem **fair**, fit preconceived notions, old paradigms, and unstated phantom criteria that no longer apply.

---

A HABIT THAT SPOILS R&D CREATIVE THINKING: We make **unwarranted assumptions** about problems and do not check them out. We stay **stuck** in old paradigms and old mind funnels.

---

You trigger your mind funnels and paradigms by words, remote associations, visual impressions, ideas, etc. They keep you glued to the past. Connecting new problems with old mind funnels and paradigms produces the same time worn solutions and spoils creative thinking, the closed mind syndrome.

You easily get locked into an old, ineffectual mind funnel or paradigm, because you maintain it with old ideas and traditions, not by current success. Shift into new mind funnels by shifting paradigms.

A HABIT THAT SPOILS R&D CREATIVE THINKING: The **quick fix** depends on accepting the first adequate solution to a problem, thereby denying your creative ability to find a better solution.

To avoid the quick fix, set a quota for three to five different ideas before choosing a solution. Or non-evaluatively list all the ideas you can think of in a three-minute brainstorming session.

A HABIT THAT SPOILS R&D CREATIVE THINKING: One habit based on the quick fix includes **rushing** to generate solutions before carefully **defining the problem** (or examining alternative mind funnels) to make sure you work on the right problem. You use **old** paradigms instead of new ones.

Old mind funnels and paradigms distort current reality and produce an inability to even see other alternatives. They lead to low quality solutions if you use the wrong funnel. Since they get bigger each time they successfully solve a problem, they diverge from reality as time passes. We refer to successful mind funnels as 'perspectives,' while we refer to unsuccessful mind funnels as 'ruts.'

A HABIT THAT SPOILS R&D CREATIVE THINKING: We do not **search** a single mind funnel for the entire range of possible new ideas.

A HABIT THAT SPOILS R&D CREATIVE THINKING: We do not **explore** new ideas for additional new mind funnels and new paradigms.

Later in this book, we will use advanced creative thinking techniques to alter these habits, get into different mind funnels, and shift to new paradigms by defining problems and listing many 'how-to' problem statements. We will then use idea-generating techniques. We will use techniques to select and combine ideas into trigger-proposals, and then into quality solutions.

## MIND FUNNELS AND "HALF OF EIGHT"

How many ways do you think you can represent "half of eight?"

Write down the number here (       ).

Now list all the ways you can think of to represent "half of eight." Spend at least five to ten minutes before you move on.

People in my creative thinking workshops have represented "half of eight" in the following ways:

### • MATHEMATICAL MIND FUNNELS

(1 x 4), (2 x 2), (3 x 1.25), (4 x 1), etc.

($2^2$ ), (square root of 16), (2 times the square root of 4), (4 times the square root of 1), etc.

(1+3), (2+2, (3+1, (5-1), (6-2), etc.

(8 divided by 2), (12 divided by 3), (16 divided by 4), etc.

### • MIND FUNNELS THAT SLICE "8" IN HALF

Slice the 8 horizontally in half to produce o and o, which are the top and bottom halves of 8.

Slice the 8 vertically into the left and right halves of 8.

Halve the 8 in all directions leading to an infinity of distorted answers. Indeed, you might halve all diagrams of eight in all directions, including eight, VIII, 4+4, and other representations of eight.

Thus:          EIGHT

### • MIND FUNNELS THAT WRITE 'FOUR' IN DIFFERENT WAYS

4; four; IV; IIII; etc.

Ideographs that write 'four' in Chinese, Japanese, Sanskrit, Arabic, Hindu, ancient Egyptian, etc. Braille (for the blind).

## • MIND FUNNELS USING CODES FOR FOUR:

100 (represents 4 in binary numbers); 11 (represents 4 in ternary numbers), etc.

Morse or semaphore code.

Deaf sign language.

Boat pennant representing four.

Sign of Four (see the Sherlock Holmes story).

500 (1000 is the binary number for 8; one-half of this is 500); also 10 and 00 (cutting 1000 in half vertically).

## • OTHER MIND FUNNELS:

Show four fingers (what a four-year old does when asked his or her age).

7:30 (the German, halb acht).

Hit the ground four times with his hoof (what Clever Hans, the horse, did).

A friend of mine suggested the following mind funnels for which we found no solutions: "half of eight using our senses: taste, touch, smell, sight and sound." Later, I thought of one involving sound. The telephone company has a universal frequency for 'four' that represents "half of eight" in sound all over the world.

Now imagine you sit in my creative thinking workshop and you only heard me say: "List all the ways to represent half of 8." I do not write it, just said it. Would you get into the following mind funnel ... half of ATE. If you did, how would you use it? Would you halve 'ATE' in all directions. Would you write "hungry" or draw a half eaten apple or an apple pie cut into 4 pieces?

## LEARN ABOUT YOUR MIND FUNNELS FROM "HALF OF EIGHT/ATE"

You can learn a lot from half of eight/ate.

First: Numerous and diverse mind funnels exist for all R&D problems, even one as seemingly simple as half of eight, and certainly for the many R&D problems we take for granted as we attempt to solve them at work. Yet, we blithely continue the quick fix, ignoring rich possibilities. You say you don't have time, another spoiler of creative thinking.

---

A HABIT THAT SPOILS R&D CREATIVE THINKING: We do not spend enough **time** shifting paradigms. You do not allot enough time to explore different mind funnels, to define problems relentlessly, and to avoid the quick fix.

---

Second: In my creative thinking workshop, people suggest many solutions to the "half of 8" problem; yet each person discovers only a few. That lesson clarifies: one of the reasons to use creativity teams includes the sharing of mind funnels to shift paradigms. Each person has unique knowledge and experience, and therefore his or her mind funnels provide unique and valuable viewpoints. Later, we will examine techniques to ensure effective sharing of mind funnels, paradigms, and perspectives in teams.

Third: Do not rush when solving problems. A hasty, early choice cuts down on quality possibilities. Creative thinking takes time and often means communicating with other people to discover new mind funnels and paradigms.

---

### • CHAPTER 7 •

### NINE MAGIC DOTS and R&D

" R&D people experience a great deal of fun when creative.

Not so much HA-HA fun as A-HA fun."

---

Many people who know the following puzzle think there it has only one answer. If you agree, prepare for a surprise. If you have seen it before, solve it anyway; shift the paradigm and generate another quite different solution.

Here are nine magic dots.

```
O     O     O

O     O     O

O     O     O
```

The problem: Draw **four** connecting straight lines that will touch all nine dots only once without lifting your pen or pencil from the paper.

If you have done it before, find a totally different solution. Spend at least five minutes before reading further.....

If you did not solve the problem, what do you see when you look at the nine dots? If you see a square, or two triangles, or some geometric figure, then you probably blocked yourself by assuming boundaries that do not exist and staying within mind funnels that kept you inside the lines.

---

A HABIT THAT SPOILS R&D CREATIVE THINKING: We often **assume boundaries** that may not exist. We stay within the lines. We think within the rules. We use unstated, phantom criteria. We use past company policies and attitudes as guidelines on how to do things without checking it out with others. We don't shift paradigms without permission.

---

You can connect these nine dots with **four** straight lines by moving outside the boundaries as shown here:

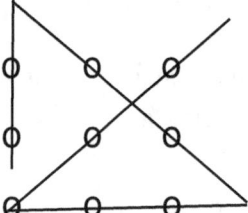

Do you like this answer? Do you think it elegant, the only one possible? Actually, the biggest assumed boundary of this problem comprises the unwarranted assumption that only one answer exists. In fact, you can find dozens of completely different answers to this problem; the one above constitutes a quick fix, the first adequate answer! How can we shift the paradigm and find the others?

We will use a very important, advanced creative thinking procedure I call '**forced-withdrawal**,' in which we forget the original problem and work to solve a distant version of it. In that way we may find new paradigms, new perspectives, and new solutions. The first forced-withdrawal we shall consider is...

Here are the same nine magic dots.

O    O    O

O    O    O

O    O    O

The problem: This time use **three** connecting straight lines that touch each dot only once. If you don't solve it, try to discover the mind funnels, assumed boundaries, unwarranted assumptions, and unstated criteria that block you.

First: What do you see when you look at the nine dots? I hope you kicked the habit of seeing a square or some other geometric figure. This time one block comes from seeing the nine dots on a piece of paper. For some solutions to this three-line problem, you need to perceive the nine dots as existing in space, because the lines will leave the paper.

Second: Did you assume that the lines must go through the center of the dots? This unwarranted assumption also blocks you.

Third: How do you define a dot? In school, I learned that a dot represents a point in space with no dimension: without length, width or height. Those circles I call dots have length and width. Is that fair? Well, in real life, dots have length and width, and come in all sizes. On billboards, dots grow to the size of your head, and on clown costumes, polka dots fit the size of your shoe. So include reality in your definition of dots, lest you fall victim to another spoiler of creative thinking.

---

A HABIT THAT SPOILS R&D CREATIVE THINKING: We use **restricted definitions** that limit our mind funnels and our creative thinking. We stay stuck in old paradigms.

---

With expanded boundaries, clarified assumptions, and unrestricted definitions, we can solve the 9 dot, 3 line problem in this way:

**Go off the paper, if necessary**

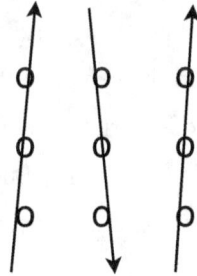

**Go off the paper, if necessary**

The first line touches the side of the first dot tangentially, passes through the center of the second dot, and touches the side of the third dot tangentially. Extend the line as far as necessary, even off the paper, so the second line can do the same to the middle row of dots, and similarly for the last line and the third row of dots.

There's another solution based on non-Euclidean geometry which postulates that parallel lines meet at infinity. Using this mind funnel, the answer consists of three parallel lines, each of which touches a different row of dots, and then all three lines connect at infinity, a neat paradigm shift.

A TRUE STORY: In a creative thinking workshop, one participant said she discovered this solution based on non-Euclidean geometry, but discarded it because she thought it unfair. This solution lay outside her comfort zone.

A HABIT THAT SPOILS R&D CREATIVE THINKING: We only express 'fair' ideas, even before we select one. Don't let **fairness** spoil creative thinking in your head.)

Here's another forced-withdrawal with the same nine magic dots.

        o   o   o

        o   o   o

        o   o   o

The problem: Use **two** connecting straight lines that touch each dot only once.

Think it impossible? Check your assumed boundaries, unwarranted assumptions, unstated criteria, restricted definitions, mind funnels, and paradigms.

One block to this problem lies in your restricted definition of a line. In school, teachers define a line as a series of connected points that have only one dimension, length. In real life, lines have width. Look at traffic lines in the center of the road or lines of buses approaching an intersection. Again your habit of restricting definitions blocks you and led to the unwarranted assumption that you could use only thin lines.

Here's one answer to the 9 dot, 2 line problem: A wide line and a narrow line!

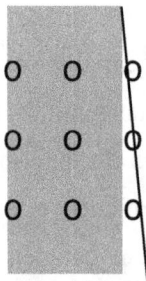

Try one last **forced-withdrawal** with the same nine magic dots.

This time use only **one** straight line that touches them all. Find at least 15 answers before you continue reading.

Actually hundreds of acceptable solutions exist. The few solutions here will trigger new paradigms and mind funnels, and whet your appetite for more.

• Use one wide line that touches each dot.

• Run a large 3-dimensional line down through the nine dots from above so it passes through the paper, and touches each dot.

• Fold the paper so you can draw one line that touches each dot. (Did you **assume** you could not fold the paper?)

• Cut the paper so each dot is on a separate piece. Line up the pieces so one line touches each dot. (Did you **assume** you couldn't cut up the paper?)

• Twist the paper into a cone and draw a straight line that spirals around the surface of the cone and touches all nine dots. (Did you **assume** you couldn't twist the paper into a cone?)

• Put the paper with the nine dots on the equator of the earth and carefully draw a straight line that circles the earth enough times so it touches each dot. Or, put the paper on the edge of the universe and have your straight line circle the universe until it touches each dot. (Did you **assume** you could not use fantasy? Note we expanded our mind funnels from nine dots in a box to the edge of the universe).

• Write "ONE" over the top row of dots, "STRAIGHT" over the middle row of dots, and "LINE" over the bottom row of dots. You touched the dots with the words: "ONE STRAIGHT LINE." (Did you **assume** you could not use words?).

• Draw the line on the thin edge of the paper. View the nine dots through this side line.

• Move a straight line, like the windshield wiper on a car, and touch all dots. (Did you **assume** you couldn't move the line, or that the line had to touch all the dots at the same time?)

• Cut the line into 1,000 pieces and sprinkle it over the nine dots touching them all. (Did you **assume** you couldn't cut up the line?)

• Cut the paper so there is one dot on each piece of paper. Line up the dots on top of each other. Push a pencil through all the dots. You not only touched all the dots with one straight line, but you also annihilated the dots and the problem. About time, I'd say.

• Wait. Here's another solution to jolt your mind. Imagine you sit in my creative thinking workshop and I merely say: "Touch each dot with only one straight line." Not write it, so you could see the spelling, but you only hear the words. One solution: bring in the king of beasts (or a picture of one) and cover the nine dots with one straight 'lion.' Or how about nine people named Dot eaten by one straight lion?

• I can't resist even more bizarre solutions. Change the dots into clothespins and hang them on one straight clothesline. (Did you assume that you could not convert the dots or the line into something else?)

• Or change the dots into tennis balls and play tennis with them until all have touched the tennis net made from one straight line. Or change the line into the shadow of a sun dial so it will eventually touch all the dots as the sun moves across the sky. Or convert the straight line into a sunbeam and use a glass prism to break it up into many colored lines that touch all nine dots. Had enough?

In my workshops on creative thinking, I always hear new and different solutions from the participants. See how many new solutions you can discover. Send me some. I always enjoy more.

## NINE MAGIC DOTS CAN HELP THE CREATIVE CLIMATE OF YOUR MIND

You should realize by now that this puzzle represents a metaphor for problems at work. You can learn a lot from these nine magic dots.

First: You may recall I stated my first answer for the nine-dot, four-line problem represented a quick fix, the first adequate answer. Here's why. You can apply many answers from the one-line problem to the four-line problem. How? Add 3 straight lines to the one straight line, thus, _____ $\wedge$. For example, use these four lines like a windshield wiper to solve the four-line problem. Did you **assume** all four straight lines had to touch a dot? Another unwarranted assumption! See below for many other unwarranted assumptions about the nine-dot problem.

Thus, we learn that even an excellent first answer mimics a quick fix, especially when people stop the creative thrust and fervor too soon. Note how effectively **forced-withdrawal** avoids the quick fix. We shall use this procedure again in Chapter 13.

Second: No one person in my workshops generates many of the solutions to the one-line problem. The sharing of mind funnels and paradigms yielded numerous solutions, an important reason to use teams to solve problems.

Third: Each team in my workshop always generates unique solutions not generated by any other team. We would lose those solutions had that team not existed. The lesson: Use more than one team to solve the same problem to produce a diversity of ideas and quality solutions.

Fourth: In my creative thinking workshop, some participants claim one solution is the best, or better than others. My response: you cannot tell until you know the criteria for an effective solution. And I never state any criteria. Indeed, if you know the criteria before you generate ideas, you box in your imagination and evaluate each idea mentally against the criteria. In this way, you lose many ideas, especially trigger-ideas, and spoil creative thinking for quality solutions.

You don't generate good or bad ideas. Ideas either fit or don't fit your stated criteria.

---

A HABIT THAT SPOILS R&D CREATIVE THINKING: We list **criteria** to choose ideas before we list ideas and thereby limit idea generation and diminish quality solutions.

---

Do not make the criteria known before idea generation. Resist comparing ideas to the criteria and your imagination will soar. See Chapter 11 for what to do with prematurely presented criteria or phantom criteria, those that exist only in your mind.

Fifth: Unwarranted assumptions block us from most of the imaginative solutions to the nine-dot problem. Some of these **unwarranted assumptions** include:

• There's only one right answer.

• Stay inside the square or the edge of the paper.

• The lines must go through the center of the dots, and touch at least one dot.

• Use Euclidian geometry only.

• Use only the mathematical definition of a dot or a line.

• Attach the lines only at their tips.

• Lines are not words spelled 'lion.'

• Lines are not phony sounding cliches.

• Don't move or expand the dots; or put them on a windshield; or convert them into popcorn, tennis balls, clothes lines. fishing hooks, or the names of people.

• Don't widen or cut up the lines; or connect them in your mind; or connect them using other lines; or connect them along their lengths; or convert them in your mind into pencils, tennis nets, windshield wipers, clothes lines, fishing lines, fashion lines, movie lines, etc.

• Don't cut the paper; or view it from the side; or put it on the equator or the edge of the universe; or fold it; or twist it into a cone or cylinder.

• Spell 'nine' only as 'n-i-n-e,' and not as 'n-e-i-n.'

• Lines are rigid and permanent, not flexible, elastic, and temporary.

• Solutions must seem 'fair' and fit our unstated, phantom criteria.

• All lines must touch the dots at the same time.

Learn to detect unwarranted assumptions when approaching any problem at work. A little effort pays off greatly as your creative thinking soars.

---

**• CHAPTER 8 •**

**TECHNIQUES TO HELP THE SHARING OF R&D IDEAS AT WORK**

---

A TRUE STORY: At one workshop, a brand manager in a Fortune-500 company told me he had learned creative thinking techniques like brainstorming before, but never the habits that affect the creative climate with clarity. He claimed a positive atmosphere was missing in his team, and that negative habits spoiled their creative thinking all the time. As he said: "Creative thinking techniques by themselves are not enough. A creative atmosphere is essential."

Most people in my creative thinking workshops make mostly negative comments when presented with a new idea. Thus, when given free choice, they make a negative comment under the guise of honest criticism, devil's advocate, or constructive criticism. Indeed, quick negative criticism commonly inflicts our society.

---

A HABIT THAT SPOILS R&D CREATIVE THINKING: We respond to new ideas with **quick negative criticism** and a **habitual automatic NO** that usually maims or kills new ideas and spoils creative thinking.

---

Criticism spoils creative thinking. Only the toughest R&D risk takers will volunteer to share the first-stage, half-baked ideas that most of us have. Successful creative people, who have written about their creative thinking, agree that quick negative criticism has a devastating effect on new ideas. Albert Einstein made this point in his autobiography.

Of all the ways to spoil creative thinking during R&D problem solving, quick negative criticism heads the list. Still, you have to give honest opinions about new ideas. Some ways to do this without spoiling creative thinking and stifling people's desire to present new ideas follows.

Suppose a person in R&D brings you an idea they like very much. How should you respond? Very carefully, I hope.

---

First, you should be thinking that whatever the idea's flaws, you need to trust this person. Consider that the new idea has merit. After all, its proposer thinks so.

Consider also that you want this person and other people to bring you ideas and proposals in the future, and you do not want to discourage this.

Also, you do not want this person to leave feeling resentful, as you have done, when people reject your ideas.

Finally, you want this person to tell you about this idea without feeling defensive, or under pressure. Abolish the hot seat antagonistic to a positive creative climate. Given all this, you do not say: "That's a lousy idea."

## USE IDEA-HELPING TECHNIQUES TO STIMULATE IDEA SHARING

How can you help the submission of new ideas? In my creative thinking workshop I recommend three techniques.

The first: '**I.P.N.C.**' Your initial comments indicate Interest (I) in the idea, and in what the proposer thinks about it, followed by all the Positive (P) comments you can muster. Then state your Negative (N) comments as **concerns**. Finally, indicate Curiosity (C) about the idea and ask how you can help the person deal with your concerns. Note that I.P.N.C. cloaks your negative comments with help and encouragement, an invitation to discuss the idea further now, and an invitation to hear new ideas in the future. Avoid sarcastic, jeering comments.

Let's see how this might work. What do you say when someone presents a new idea they like? You could use I.P.N.C. to open the discussion.

= "That's a very interesting (I) idea.

- I like it (P) because it ...

- I am a little concerned (N) that...

- I wonder (C) why you like it? How will you use it? How can I help?"

Notice that you have pointed a fatal flaw in a way to help the other person feel encouraged. People often ask why not start with direct questions. In my experience, and from what others tell me, many people perceive immediate quick questioning as a negative response that puts them on the defensive, and lowers their willingness to propose new ideas again.

The person can now respond to your I.P.N.C. comments in several ways. He or she might say: "You're right. I had not noticed that. Thanks for telling me. I'll change the concept," and leave, hopefully feeling encouraged, and, at least thinking you took his or her ideas seriously.

Now aren't you glad you didn't say "That's a lousy idea" when you first saw it? Of course, it might have been easier if the other person had made it clearer. This leads us to another habit that spoils R&D creative thinking.

---

A HABIT THAT SPOILS R&D CREATIVE THINKING: We expect sellers of ideas to present their ideas in **perfect** form. No half-developed ideas for us. Every "i" dotted, every "t" crossed, every concept clear, every label and term used correctly, and no errors of spelling or grammar.

---

In other words, we expect them to help us before we help their idea. Beware. Creative ideas rarely appear in perfect form, and negative consequences occur if you insist on this. People will expend valuable time and effort toward perfectionism. Besides, few of us have training in selling ideas anyway. See Chapter 20 for help in selling proposals.

A second approach I recommend in my creative thinking workshop to help submission of new ideas: **'Yes-If**.' Curb your automatic NO and say "Yes, if..." describing the conditions needed to get from NO to a conditional YES. Watch the climate change from a negative one spoiling creative thinking to a very positive idea-helping climate.

---

A TRUE STORY: At the end of one creative thinking workshop, an executive who tried "yes-if" for the first time said he noticed that "yes-if" not only kept him away from his automatic No, but led him to listen to the new ideas presented to him very intently to discover the ways to convert a NO to a conditional YES. As he put it: "Yes-if converted me into a collaborator to help a new idea get going, rather than staying a judge."

---

---

A TRUE STORY" A plant manager wrote me after a workshop that "Yes-if turned me into a better listener, one who is more empathetic with his people, better understood, and now more predictable. My effectiveness improved by leaps and bounds!"

---

Let's see how you might use Yes-if.

- You could say "YES," there are many interesting and useful features about this idea,

- "IF, we can improve..."

- Clear, supportive, and crisp. Very useful when hearing ideas on the run.

A third approach I like: 'Idea Improvement Analysis (I.I.A.).' Rate each aspect of the idea on a scale of 1 to 10. Consider those elements that fall below 3 or 4 as snags you need to overcome, rather than reasons to reject the idea. This procedure helps develop new ideas rather than focusing on negative aspects that spoil ideas and egos.

Here's how Idea Improvement Analysis (I.A.A.) works:

- "I rate the ... about a 'nine' and the ... about a 9.5, especially since ... I rate the ... very high, because it could be very useful for ...

I rate its ... very low, about a '1.' How can we overcome this difficulty?"

I.A.A. Presents a very detailed analysis of a proposal and shows genuine interest.

If none of these approaches seem comfortable you might use "What's Good About It" and state three positive statements first.

Or you could pretend your boss presents any idea you hear and make up your own idea-helping approach. Your attitude counts a great deal when you are helping ideas. Increased idea sharing and better idea improvement will result.

Other statements to encourage new ides include:

• The value of that idea is....

• That seems like a useful idea. Can we build on it?

• A good start. Can we help it?

• It seems we are getting somewhere.

• Describe that in more detail. Tell me more about it.

• How can we make this work?

• I would be interested in what you have to say.

• Tell me what you are thinking.

• I like your idea. Let's see how we can get over this snag.

• That idea has value. Let's get the bugs out.

---

• You may very well be right. Still, let us look at it another way.

## QUICK SPOILERS OF R&D CREATIVE THINKING

The more polite alternatives to I.P.N.C., Yes-if, and I.I.A. include the following quick creative thinking **spoilers**.

• It's already been patented.

• We've never done it that way before... We've tried that before.

• If it ain't broke, don't fix it.

• Once you define the problem properly, the solution seems obvious (so we don't need to generate many ideas).

• We don't need any more new ideas around here; what we need are more doers, implementers, product champions, etc.

• The problem with that idea is...

• It's not in the budget ... or We haven't the personnel.

• What will the customers (unions, top management) think?

• Somebody would have suggested it before if it were any good.

• We're too small for that ... or We're too big for that.

• We have too many projects already.

• It has been the same for twenty years so it must be good.

• I just know it won't work.

• That's not our problem... or That's not our responsibility.

• Engineering can't do it...Production won't accept it.

• You'll never sell that to management.

• Why something new now? Our sales are still going up.

• Not in the plan... No regulations covering it.

Stop using these quick R&D spoilers; they have many detrimental effects.

First, many R&D people suppress and stop expressing their ideas.

Second, creative R&D people stop being creative on the job and save their creative thinking for the weekend. They become what I call the "weekend creatives." Here's what one participant from a Fortune-500 company wrote: "I have found it necessary to turn to off-plant community activities and hobbies to fulfill the urge to be creative." He proved this by proudly showing me his creative output in his home.

Third, some R&D people become defensive, apologetic, and shoot-down their own ideas before anyone else does.

Finally, some R&D people tend to do things the same old safe, complacent way instead of taking risks, exploring new mind funnels, shifting paradigms, and developing half-baked, half-developed, bizarre ideas into winners. Not all of these pan out, of course, but unless you deliberately and relentlessly help develop new ideas, they perish, and complacency takes over. I do not remember who said it, but if you want to be around butterflies, you have to generously help many caterpillars.

Help new ideas with I.P.N.C., Yes-if, and I.I.A. Or pretend every idea you hear comes from your boss.

## ASSERT THAT YOU WANT OTHERS TO HELP YOUR R&D IDEA

When a work group quickly shoots down ideas, many R&D people get defensive and start shooting down their own ideas first, if they actually muster up the courage to share them at all. Some ways they do it includes:

• This may not work, but...

• You'll probably laugh, but...

• It might be a dead end, but...

• I'm no genius, but...

Stop shooting down your own ideas with such apologetic phrases. Instead, assert that you want others in R&D to help you develop your new idea further. Take responsibility. Change the creative climate around you.

---

A HABIT THAT SPOILS R&D CREATIVE THINKING: We don't **object** when R&D people stifle our ideas. We allow other people to shoot down our ideas and spoil our creative thinking. We do the same to our own ideas. Stop doing this. Take responsibility for the creative atmosphere around you.

---

## TAKE RESPONSIBILITY FOR THE R&D CREATIVE CLIMATE

How responsible are you for the creative thinking of others at work? A lot? A little? Not at all? Actually, 100%.

Why 100% responsible? Because you can say "NO" so easily. Your automatic No devastates the creative thinking of others. Use I.P.N.C., Yes-If, I.A.A., and other techniques to create a creative climate for others. Pretend all the ideas you hear come from your boss.

How responsible are you for your own creative thinking? A little? A lot? Actually, 100%.

Why? Because you can assert to R&D people with a large automatic No and quick negative criticism. Ask them to help you develop your idea, not kill it. Assert to create a creative climate around you.

Accept 100% responsibility for the creative climate at work. You can use I.P.N.C. and Yes-If to help others. You can assert to help yourself. You can change a negative climate to a positive one by your own actions!

To help creative thinking in R&D, squelch criticism, not new ideas. A new idea is like a brown, ugly seed. You do not know whether it will grow into a lovely flower or a common weed until you plant it and nurture it. A newly formed idea resembles this too, in that it is half-developed ... or half-baked; what you call it depends on whether it is your idea, or whether you like it.

People find new ideas in R&D uncertain. You do not know the direction the idea will take or whether they will get there. Many mistakes will occur. Surrounded by high risk, no one can predict the future and prove in advance any new idea will succeed. Because R&D people find new ideas unpredictable and hard to develop, people find them easy to reject.

---

A HABIT THAT SPOILS R&D CREATIVE THINKING: We discourage & **squelch** new ideas, especially bizarre ideas.

---

Remember the story in Chapter 5 about turning the bears and the pots of honey into the down blast of a helicopter. Use bizarre trigger-ideas as starters to spark more useful ideas in R&D.

## SUMMARY: FOSTER THE CREATIVE CLIMATE IN YOUR R&D WORKPLACE

By creative climate, I mean the attitudes and behaviors that lead to the freer use of everyone's ideas during problem solving. This includes using helping and reinforcing responses to new ideas in R&D, the willingness to examine and explore different points of view, the deliberate searching for new connections between facts, beliefs, and ideas to create quality solutions.

You can foster such a positive climate in R&D by looking for the good in ideas before concentrating on what is bad by first using I.P.N.C., Yes-If, and I.I.A. as described above. You can stimulate others to do likewise. You can be eager to hear new ideas and to find whatever is good and useful in them. You can eliminate the times you and others take automatic pot shots at each other's ideas.

The quick automatic-NO impedes a creative climate for solving problems in R&D. You can help by encouraging positive climate factors. You can set the example, the tone, and the mood for everyone else. A positive creative climate is a rarity, rather than commonplace, so it takes much courage to stay creative and to express new ideas. And it takes courage to help another person's idea that seems worthy of your quick negative criticism. It takes courage to defer judgement of ideas during problem solving.

You can become the prime mover toward a positive creative climate. You need to stay optimistic. You need to convince others by modeling the behaviors you want to encourage. You need to understand the great disadvantages of the quick automatic NO response, and you must help the what's-good-about-it approach until it becomes the habitual response of other people around you.

Creative thinking support leads to more creative thinking support. Lack of creative thinking support leads to more lack of creative thinking support.

---

• CHAPTER 9 •

## AUTOMATIC WRITING HELPS
## THE CREATIVE CLIMATE IN YOUR MIND

---

The creative climate in your mind has profound effects on your creative output. Your paradigms, beliefs and thoughts propel your behavior into creative or not-so-creative activities. For example, if you believe you can't think creatively, then you won't. And what's worse, you won't make the effort to learn how.

Here's another. If you believe in certain assumed boundaries and unwarranted assumptions, or adhere to certain unstated criteria and an unrealistic sense of fairness in your mind, then your belief system and paradigms will keep you stuck in limited mind funnels and spoil your creative thinking.

And here's a crushing example. If you have a highly developed habitual automatic No and a fondness for quick negative criticism, not only will you spoil everyone else's creative thinking, but you will cut down on your own R&D creative output. Your creative thinking will be mired in excessive gloom and prophecies of failure, even though you can think very creatively, pulling negative comments out of the blue.

'**Automatic writing**' provides one antidote to an excessively quick negative mind. I learned it from William Drath at The Center For Creative Leadership.

To fully understand it, write a short essay on one of the following topics:

"What I did on my last vacation"

or

"What I plan to do on my next vacation."

Write your choice here.

Now please turn to the next page.

---

Before you start writing, please plan carefully in your mind what you want to say. Compose a well-written essay with correct grammar, full sentences, and appropriate paragraphing. In my creative thinking workshop, I would indicate that I may ask you to read your essay aloud. In other words, write a clear, orderly exposition. Start now. Stop after three minutes, so please time yourself carefully. Finish writing on another sheet of paper, if necessary.

PLEASE DO NOT READ FURTHER UNTIL YOU HAVE WRITTEN FOR THREE MINUTES.

Now write a short essay about the topic you did not choose. This time, do not plan any ideas ahead of time. Write while you think quickly.

Forget correct grammar. No complete sentences. Incomplete phrases will do. No paragraphing. Do not evaluate what you write. Let your thoughts flow directly to the paper through your pen or pencil. Do not stop writing. If you stop writing, you probably evaluate what you think.

When you have no thoughts, write something anyway. If necessary, write "I have something to write" repeatedly until your thoughts start flowing. Do not let your pen or pencil stop writing. Best of all, no one will ask you to read your essay aloud. Start now and please stop at the end of three minutes. Finish writing on another sheet of paper, if necessary.

PLEASE DO NOT READ FURTHER UNTIL YOU HAVE WRITTEN FOR THREE MINUTES.

Count the words and ideas you wrote each essay.

If you are like most people in my workshops, you will have more words and more ideas in the second essay than the first. Freeing you from evaluation and quick negative criticism short-circuits your habitual automatic No, improves the creative atmosphere in your mind, and helps you produce more creative outcomes.

Some guidelines for automatic writing follow:

• Write all thoughts.

• Do not hold back. Let it flow!

• Orderly thoughts not required.

• Correct grammar unimportant.

• Incomplete phrases fine.

• Complete sentences not necessary.

• Paragraphing not important.

• Do not evaluate.

• Bypass your hand. Be the paper. Let your thoughts flow directly to the paper.

• What you write does not have to fit the topic. That boxes you in and you measure every thought against the topic.

• If you do not write, assume you evaluate. Write "I do have something to write" until you have something to write, and then write, write, write...

I use automatic writing, a valuable Zen-like creative thinking procedure, to overcome obstacles to my creative thinking or my writing. It usually cures my writer's block. It almost always allows interesting ideas to emerge. Install a creative atmosphere in your mind to help you make remote associations to shift paradigms and solve problems creatively.

## ADVANCED AUTOMATIC WRITING

Practice an advanced version of automatic writing by placing two writing tablets next to each other. On one tablet, write automatically as described above. On the other tablet, write "I have something to write" when you find yourself blocked. To help overcome the block, switch from one tablet to other as the spirit moves you.

Another approach: write your ideas as they come to you on 3" x 5" index cards, one idea per card (See 'Idea Card' in Chapter 12). When finished, sort the cards in any order you want. Use these cards as an outline as you write automatically on the two writing tablets as described above, except now use one tablet for the finished writing you want and switch to the other tablet for random thoughts when you find yourself blocked.

The important thing: keep your hand writing automatically. Merge with the paper as you write. Become the paper and pen.

A TRUE STORY: I wrote the first draft of this book, and many subsequent added sections, using automatic writing.

# • PART 3 •

## TECHNIQUES TO SHIFT R&D PARADIGMS

Creative thinking techniques are useless unless used.

--------------------------------------------------------

## • CHAPTER 10 •

## AN EFFECTIVE PROBLEM-SOLVING SEQUENCE TO SHIFT R&D PARADIGMS

## THE THREE CREATIVE STEPS

A TRUE STORY: A manager in a Fortune-500 company asked me to lead a workshop for an R&D group that had no previous creative thinking training. I suggested two days of training on advanced techniques, and then a third day working on problems using their newly learned habits and techniques.

The 18 participants trained in three teams for two days. On the third day, each team worked on a different problem of the organization. One team did extremely well. They identified ten unexpected uses for a new technology and sent the ideas to the appropriate people in the company. The other two teams did well, but not to this level.

## AN EFFECTIVE PROBLEM-SOLVING SEQUENCE TO SHIFT R&D PARADIGMS

Study this problem-solving sequence; it works well to shift R&D paradigms and produce high quality solutions (* indicates the key creative steps).

*Step 1. Define the R&D problem by generating many problem statements.*

 Step 2. Identify criteria to select the problem statements.

 Step 3. Select the problem statements on which to focus.

*Step 4. List many ideas.*

*Step 5. Combine ideas into creative trigger-proposals.*

 Step 6. Identify the criteria to select quality solutions.

 Step 7. Convert trigger-proposals into quality solutions that meet the criteria.

 Step 8. Make action plans to implement or sell the quality solution to other people.

This sequence incorporates some important concepts essential to shift R&D paradigms and produce high quality solutions.

**First**: The three key creative steps in problem solving...

- Step 1: Define the problem;
- Step 4: List many ideas; and
- Step 5: Combine ideas into creative trigger-proposals.

Welcome bizarre and exotic trigger-ideas in each step. Use them to spark better ideas. Stay positive throughout. Let your imagination soar. Do not discard or ridicule any idea. Instead, choose what you want to use. Keep what's left for future reference or discard them by gentle neglect. The more bizarre you define a problem, the more likely your imagination will produce a paradigm shift and a practical solution that differs from past R&D approaches. Thus:

| Bizarre | | Bizarre | | Bizarre | | New and Different |
|---------|----|---------|----|----------|----|-------------------|
| Problem | => | Risky | => | Trigger- | => | Workable |
| Statements | | Ideas | | Proposals | | Solutions |

**Second**: Avoid rushing to generate solutions until you extensively **define** the problem to make sure your team works on the right problem. Do not stuff the new problem into a comfortable old mind funnel or paradigm.

R&D people who spend more time on Step 1 (defining the problem) usually produce solutions more creative than people who rush to Step 4 (generating ideas) first. This makes a great deal of sense, since jolting your mind first to pursue new directions, new paradigms, and new mind funnels ensures that you generate unusual ideas and solutions.

A HABIT THAT SPOILS R&D CREATIVE THINKING: We **rush** to generate solutions before we adequately examine and define the problem, a variation of the quick fix.

**Third**: Do not actively reject unacceptable problem statements or ideas. Just leave them behind as you chose others. They may work later.

**Fourth**: Do not identify the criteria before Step 6. If you do, you box in your creative thinking as you prematurely measure problem definitions and ideas against the criteria.

---

A HABIT THAT SPOILS R&D CREATIVE THINKING: We box in our creative thinking by **identifying criteria** before we generate problem statements or ideas.

---

**Reverse your criteria.**

Often R&D people prematurely obtain criteria from others or remember unstated, phantom criteria from previous experiences. Deal directly with such criteria or they will inhibit and scuttle your creativity. Get rid of them. Use forced-withdrawal and reverse the criteria.

1. Non-evaluatively list all given and unstated, phantom criteria. Set a quota for at least 10 to 20 phantom criteria.

2. Reverse your criteria. Distort them. Shrink them. Magnify them. Play "what if" with them. Do whatever you can to make them insignificant.

3. Later, when you finish generating ideas and trigger-proposals (see Chapter 13) you can return to realistic criteria (Step 6). Then select ideas and proposals that fit your criteria.

**Fifth**: After all that effort and fun, every solution you intend to implement deserves an action plan with detailed action steps (see Chapter 15). Proposals without a plan perch perilously close to perishing.

---

**· CHAPTER 11 ·**

**CREATIVE THINKING IN ACTION**
**STEP 1. TECHNIQUES TO SHIFT PARADIGMS AND**
**DEFINE R&D PROBLEMS CREATIVELY**.


(THE 1ST CREATIVE STEP)

---

You know many ways to define R&D problems logically. Don't abandon them. In addition, define the problem innovatively (Step 1 in the problem-solving sequence). Talk the problem out; think the problem through; everything helps shift the paradigm.

Gaining new perspectives and new paradigms while defining an R&D problem creatively can happen in many ways. Step back and pretend you are someone or something else; say, a frog, an elephant, a sea gull, a shark, or a person from another culture or another profession. Deliberately reverse and distort your view of the problem. Systematically list what you like about the problem area and what you want to improve. Answer the who, what, where, why, when, and with whom about the problem.

These approaches jolt your mind into a multitude of fresh mind funnels and new perspectives that shift paradigms for quality solutions, Above all, you need to list numerous diverse problem statements. I have provided specific directions and detailed methodology in this chapter.

---

A HABIT THAT SPOILS R&D CREATIVE THINKING: We do not use enough techniques to **shift paradigms** and define problems innovatively.

---

We often spend time solving the wrong R&D problem, so the problem appears repeatedly, never effectively identified and resolved. To end this cycle, transform and modify the starting problem statement, generate a long list of diverse problem statements before idea generation and open up a broad range of new perspectives and paradigms, a key step to achieving quality solutions in R&D.

---

Work in teams, if possible. Different perceptions and paradigms more likely surface in teams because of the unique experiences and mind funnels each R&D person brings to the process.

The problem statement will dictate the ideas you generate later. So focus on more than one problem statement to keep you from overlooking the obvious or committing to an inappropriate solution too soon. Make sure you solve the **real** R&D problem.

## TURN ON YOUR "HOW-TO" THINKING

Start your problem statements with 'How to' to keep you from getting into solutions too soon. The techniques described below provide help. Spend hours, even days, writing 'How-to' problem statements if the problem warrants it.

### • 'How-To' R&D Problem Statements.

By listing many problem statements, you search for new, fresh R&D paradigms, mind funnels and unexpected definitions of the problem. Such statements provide a broad-based pyramid upon which to later generate ideas and develop solutions. Do not turn the pyramid upside down by rushing to generate solutions. Shift the paradigm, define the problem, first.

---

A HABIT THAT SPOILS R&D CREATIVE THINKING: We turn the problem pyramid **upside down** and rush to generate solutions before extensively defining the problem and establishing a good base on which to generate ideas.

---

Write your starting problem statement:

How to...


Define your problem by completing the following:

| | | |
|---|---|---|
| How to... | How to start... | How to establish... |
| How to gain... | How to expand... | How to afford... |
| How to improve... | How to amplify... | How to increase... |
| How to change... | How to build... | How to reject... |
| How to add... | How to manage... | How to persuade... |
| How to fix... | How to commence... | How to enrich... |
| How to minimize... | How to enlarge... | How to provoke... |
| How to accomplish... | How to establish... | How to encourage... |
| How to enhance... | How to succeed... | How to invent... |
| How to cope with... | How to attempt... | How to create... |
| How to restore... | How to adapt... | How to innovate... |
| How to do away with... | How to originate... | How to enjoy... |
| How to produce... | How to schedule... | How to appreciate... |
| How to exceed... | How to grow... | How to achieve... |
| How to reduce... | How to maximize... | How to tell... |
| How to deliver... | How to learn to... | How to share... |
| How to make the | How to perform... | How to distribute... |
| best use of... | How to flourish... | How to assemble... |
| How to handle... | How to make... | How to reverse... |
| How to eliminate... | How to end... | How to twist... |
| How to change... | How to disclose... | How to blend... |
| How to develop... | How to restore... | How to combine... |
| How to control... | How to destroy... | How to transpose... |
| How to launch... | How to inspire... | How to interchange... |
| How to alter... | How to motivate... | How to substitute... |
| How to switch... | How to deal with... | How to rearrange... |
| How to admire... | How to lose... | How to use... |
| How to begin... | How to conquer... | How to subtract... |
| How to revive... | How to modify... | How to attempt.... |
| How to upgrade... | How to challenge... | |
| How to arrange... | How to reward... | |
| How to enrich... | How to satisfy... | |

**• Analogies and Metaphors.**

Analogies and metaphors help you think creatively in new places. For example, if you want to improve paint, start by listing the characteristics and properties of paint. Among these properties: paint provides a protective boundary. Identify other places where a protective boundary exists, list its properties, and force combinations to improve paint. Some examples follow:

• In plants: the boundary between woods and pasture depends on soil, nutrients, and cultivation. This leads to thoughts on how to make a paint that has different colors on different surfaces. In addition, fungus on the bark of trees indicates dead wood, triggering how to make paint that signals rot underneath.

• In the animal kingdom: octopus and squid skin change color with different moods, triggering thoughts on how to make a paint that will change color with the room occupant's mood. Also, animal skin has a protective boundary that can self-heal and self-cleanse, triggering how to make paint self-repair and self-cleanse.

• In physics: charges at boundaries repel, triggering thoughts on how to make a paint that repels dirt.

• In an artist's painting: the boundary between objects results from different colors and pigments, triggering how to make a paint that produces a picture when spread on the wall.

• A book: its covers trigger thoughts to make a paint that opens to reveal what it covers and then closes to hide.

Recycle this procedure using another characteristic or property of paint to shift paradigms. Expect good results.


**• The R&D Problem's Essence.**

Knowing a problem well unfortunately means you have a myriad of pictures in your mind that spoil new thinking. To avoid these old pictures, use this version of forced withdrawal to work on the problem indirectly. Start with the 'essence' of the problem, the action verb that captures the main activity.

For example: the essence (or action verb) of an auto jack is lifting things; a wheelbarrow is transporting things; walking on water is floating things or freezing water; a bullet proof vest is impenetrability; reuse of cans and bottles is recycling things;

improving the can opener is opening things; keeping food from spoiling is preserving things.

So instead of starting with how to improve the can opener, a creativity team first discussed ways to open things using analogies and metaphors from industry, animals, plants, other cultures, etc. What happened? They discussed:

- squeezing the base of a dog's mouth so it will open

- a clam relaxes a muscle so tension on the back hinge of the shell forces the clam open;

- as peas ripen, the tough covering develops a weak seam and the pea pod opens.

The team forced combinations between the weak seam of the pea pod and opening cans. This did not lead to an improved can opener, as they originally intended, but it did lead to opening cans by pulling a weak seam, a common way to open most cans now.

This example illustrates nonlinear creative thinking described in Chapter 5. It looks like this:

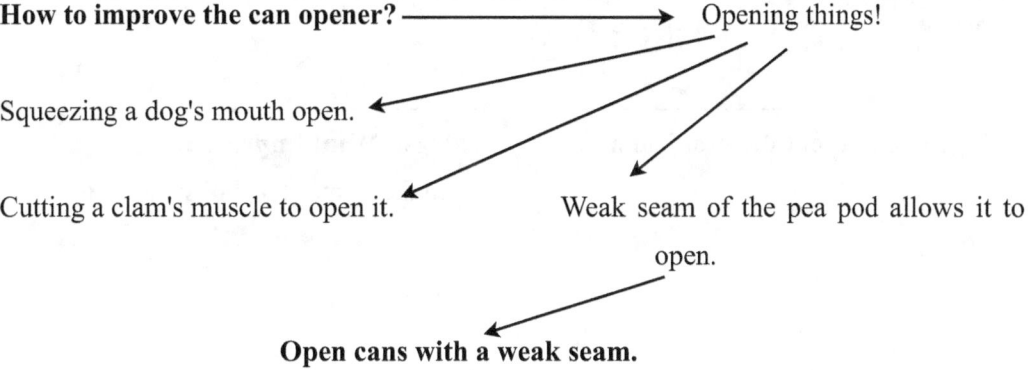

**How to improve the can opener?** ⟶ Opening things!

Squeezing a dog's mouth open.

Cutting a clam's muscle to open it.      Weak seam of the pea pod allows it to open.

**Open cans with a weak seam.**

**• Targeted Analogies and Metaphors based on the R&D Problem's Essence.**

Combine the two previous techniques into a truly exceptional approach that shifts paradigms and opens a flood of new approaches.

1. Forget about the problem statement and deal only with the action verb that captures the essence of the problem.

2. Generate examples of the problem's essence as metaphors and analogies from the plant and animal world; industry and government; professions; other countries; ethnic and religious groups; the historical past; exotic other-world places; the wild west; etc.

3. Choose one example and list detailed characteristics, attributes, and properties of the example.

4. Force combinations between these attributes and the problem statement(s) to provide exotic, bizarre ideas.

5. Turn each bizarre idea into realistic and workable ideas to solve the original R&D problem. One way to do this: non-evaluatively list each characteristic, attribute, and property of the bizarre idea and force combinations between these and the problem. Be patient and you will move toward a high quality solution.

## • Like-Improve Analysis.

Look at what you like about the R&D problem area, so you won't change that, while you dissect out what's deficient and needs improvement. More logical than the problem-defining techniques described above and still very effective to shift paradigms and define problems.

Write your problem statement and fill out 'Things I Like' and 'Deficiencies' below:

---

How

to...

---

| Things I Like about the problem area. | Things I Want Improved. |
| --- | --- |
| | (Turn these into 'how to' statements.) |
| -- | How to... |
| -- | How to... |
| -- | How to... |
| -- | How to... |
| -- | How to... |

• **Reversal-Dereversal.**

Turn your R&D problem upside-down. When you get right side up again, you will face a new direction after the paradigm shifts.

Write your starting problem statement: How to...

1. Reverse the key verb of the problem statement. For example: write spoil instead of stimulate; decrease instead of increase; fail instead of succeed; etc.

2. Non-evaluatively list solutions to the reversed problem statement.

3. Dereverse by writing "how-to" in front of each solution.

4. Smooth out the wording of the new problem statement until it makes sense. Do this creatively.

Choose an appropriate new "how-to" problem statement to use during idea generation.

Restate the problem: How to …

Example of reversal-dereversal:

1. Reverse "How to <u>stimulate</u> creative thinking in R&D meetings" **into** "How to <u>spoil</u> creative thinking in meetings."

2. One way to spoil creative thinking is: Have dominating people present in the R&D meeting.

3. Dereverse this statement to: "How to stay creative with dominating R&D people present" or "How to get rid of dominating people."

4. Choose the problem statement on which you'd like to focus.

• **Reverse Assumptions.**

Uncover the unwarranted assumptions you make about your R&D problem and use them to discover new paradigms and novel 'how-to' problem statements.

1. Non-evaluatively list five to twenty assumptions you make about your problem. Include obvious assumptions you take for granted. Drop the unwarranted assumption that you can do this task easily. Not so.

2. Reverse the meaning of each assumption.

3. Non-evaluatively force combinations between each reversed assumption and your how-to problem statement. Produce bizarre and not-so-bizarre problem statements and ideas.

4. Select, combine, change, add to, and develop new how-to problem statements.

• **Fresh Eye.**

Take on a new identity. Think about how to solve your problem as someone or something else.

---

Write your starting problem statement:

How to...

---

How would you write your problem statement if you were a...

a) Dolphin, bat, eagle, jellyfish, ion, pea pod, oak seed (choose one)

 How to...

b) Chemical engineer, mechanical engineer, Martian, artist (choose one)

   How to...

c) Biologist, chemist, secretary, banker, frog, geneticist (choose one)

   How to...

d) Architect, building contractor, carpenter, accountant, shark (choose one)

   How to...

e) Physicist, astronomer, musician, dancer, elephant, farmer (choose one)

   How to...

f) Hydraulic engineer, clothes designer, trumpet player, cougar (choose one)

   How to...

Restate your R&D problem: How to...

**• Word Substitution.**

A systematic change of a word in a problem statement transforms perspectives profoundly. New mind funnels and paradigm shifts easily occur. For example, you can transform "How to <u>get rid of</u> a dominating person"...into:

"How to <u>get rid of</u> a dominating person"

"   "  work with            "
"   "  change               "
"   "  succeed with         "
"   "  enjoy                "
"   "  do away with         "
"   "  handle               "
"   "  avoid                "
"   "  work around          "
"   "  succeed in spite of  "
"   "  get along with       "
"   "  retrain              "
"   "  negotiate with, etc. "

Note the different perspectives that occur with each verb substitution, helping you see new ways to approach this problem. Playing around with the word 'dominating' or 'person' may provide a helpful forced-withdrawal.

**• Why, Who, What, Where, When, and With Whom.**

Ask yourself questions that force you to look at a problem in a different way. More logical than some other approaches, it still leads to unexpected new perspectives.

Write your starting problem statement:

How to...

Please answer the following questions about the problem:

• Why?

• Who?

• What?

• Where?

• When?

• With whom?

Restate the problem: How to...

**• Needs, Obstacles, and Constraints.**

Force yourself to look at the R&D problem in a different way.

List needs: What do you want?

How to achieve...

How to gain...

How to...

List obstacles: What is in the way?

How to overcome...

How to get around...

How to...

List constraints: What must you accept?

How to cope with...

How to substitute for...

How to...

Write a new problem statement. How to...

**• Weaknesses of Quick-Fix Solutions.**

Get past pet solutions and old paradigms by examining their weaknesses.

1. Write your starting problem statement: How to...

2. List three quick-fix solutions and some weaknesses of each one.

3. State the problem based on what you have written:     How to...

---

A TRUE STORY: After a session in a workshop during which we extensively defined a problem creatively, a vice president of a Fortune-500 company told me he astonished himself at how many new perspectives he obtained on a problem on which he had already worked for several years in his job. He achieved important paradigm shifts using these problem-defining techniques.

---

Use problem-defining techniques described above in addition to the usual ways you use to define problems. Every problem-defining procedure helps clarify some aspect of the problem and shift paradigms toward a quality solution.

## STEP 2. IDENTIFY THE CRITERIA TO SELECT YOUR PROBLEM STATEMENT(S).

The following questions will help you identify the criteria to select problem statements on which you will focus:

Whose problem?

What kind of problem?

Marketing:

Manufacturing:

R&D:

People:

Financial:

Other:

How big?

Funds:

People:

Time:

Other resources:

Your gut feelings?

List the tangible and intangible values.

Other issues?

Check the kind of ideas you want...

( ) You want ideas that involve people.

( ) You want ideas that fit your current information.

( ) You want ideas for new plans and techniques.

( ) You want ideas for new products.

( ) You want ideas for new technology.

( ) You want ideas for new paradigms, mind funnels, perspectives, and better ways to do things.

( ) Other kinds of ideas?

Non-evaluatively list the criteria you will use to choose the problem statement.

Write five to ten more how-to problem statements.

## STEP 3. CHOOSE PROBLEM STATEMENTS ON WHICH YOU WANT TO FOCUS.

Work on more than one problem statement to make sure you do not miss anything important. Write three problem statements.

How to...

How to...

How to...

Choose the leading problem statement.

How to...

---

**• CHAPTER 12 •**
**CREATIVE R&D THINKING IN ACTION**
**STEP 4: TECHNIQUES TO**
**GENERATE NEW R&D IDEAS ABUNDANTLY**

**(THE 2ND KEY CREATIVE STEP)**

"Brainstorming doesn't accomplish enough any more ...  too old fashioned."

---

This chapter focuses on techniques to generate ideas in R&D creative thinking teams because of the larger number of diverse perspectives available through sharing in teams. Still, teams don't generate ideas, individuals do. The extent that an R&D team produces new ideas depends on the creative climate and on how the team encourages new ideas.

Adapt these idea-generating techniques to when you **work alone**. See Part 5 of this book for techniques geared to people who work alone.

**LOW TO HOT CREATIVE THINKING:**
**THE CREATIVE OUTCOME DEPENDS ON THE IDEA-GENERATING PROCEDURE YOU USE.**

Advanced creative thinking techniques don't guarantee quality solutions in R&D. Still, if you use them, you raise the probabilities that an unexpected useful outcome results.

The following list includes some outstanding idea-generating techniques with a high probability of yielding high quality creative ideas:

## LOW TO HOT CREATIVE THINKING: BOOST YOUR CREATIVITY LEVEL

CREATIVE THINKING LEVEL

IDEA GENERATING PROCEDURE

• HOTTEST LEVEL •

• A. FUTURE FANTASY:

Future pretend year

• B. FORCED COMBINATIONS:

Free word association imagery

Metaphors & Analogies

Improve bizarre trigger-ideas game

Weird to the practical idea

Idea grid; Random word triggers

Picture triggers; Quotation triggers

• C. TARGETED FREE ASSOCIATION:

Brainstorming Techniques

   Non-evaluative Listing

   Buzz groups

Brainwriting Techniques

   Automatic writing

Idea gallery

Idea card

Clustering

Brainwriting circles

• LOWEST LEVEL•

• D. UNSTRUCTURED
   FREE ASSOCIATION:

The old fashioned way that works so well

Level D above evaporates water by putting it in a large dish

Level C evaporates water by blowing on it

Level B adds heat to the process

Level A resembles a blast furnace

The effectiveness of a procedure increases enormously depending on how hot you get.

## TECHNIQUES USING UNSTRUCTURED R&D FREE ASSOCIATION

Unstructured free association involves the ordinary processes in daily on-the-job creative thinking. The remote associations sparked by unstructured free association usually produce low-creative ideas. Still, you can apply the ideas easier to the problem because they fit better.

## TECHNIQUES USING STRUCTURED R&D FREE ASSOCIATION

### • Brainstorming Techniques.

Brainstorming was invented over 80 years ago, and we have invented hundreds of newer and more effective techniques to enhance idea generation since then. Still, use brainstorming to help you flush obvious pet ideas from old paradigms and clear your mind for higher levels of creative thinking using more advanced techniques.

During brainstorming, a team of five to seven people call out ideas while a recorder writes them on a note pad or flip chart paper for all to see. It generally produces many ideas.

To increase the number of ideas expressed, Osborn, who invented brainstorming in the 1930s, devised the following guidelines:

SUSPEND JUDGMENT: No criticism. Postpone evaluation and defer judgment until later.

FREEWHEEL. The wilder the idea, the better. People find it easier to tame down than to think up.

QUANTITY: Quantity, quantity, quantity. The greater the number of ideas, the greater the likelihood of producing high quality.

CROSS FERTILIZE: Combine and piggyback ideas. Use all ideas as trigger-ideas to spark still better ideas.

A number of useful variations exist. Brainstorm for three minutes; be quiet for three minutes; brainstorm for another three minutes, etc. Or go around the team in turn, allowing people to state an idea, or to pass. Use this last variation to subdue highly vocal people and encourage quiet ones.

## • Non-Evaluative Listing.

Effective outcomes during brainstorming depend on everyone being non-evaluative to encourage R&D people to express ideas. Another reason: **evaluation forms from old information**. When we evaluate, we immerse ourselves in old paradigms.

To escape from old R&D perspectives, stay non-evaluative. I stress this point by calling the process 'Non-Evaluative Listing' and suggest the following guidelines:

- List all ideas.
- Do not discuss items or ask questions.
- Do not evaluate or make fun of suggestions. No sarcasm.
- Ignore repetition. Write ideas down again without comment.
- The recorder and the team does not have to like or understand the items listed. Questions for clarification comes later.
- Defer judgment and postpone evaluations until later.
- Keep the process moving.

These guidelines resemble the Zen-like automatic writing principles in Chapter 9.

I start an R&D idea generating session with non-evaluative listing (brainstorming) because it records everyone's pet ideas and participants stop worrying they will lose these ideas. Also, it flushes out and clears the mind of obvious solutions and makes room so advanced creative thinking techniques work even better. But if you want to produce a high quality solution, non-evaluative thinking (or brainstorming) won't get you there. Consider it, at best, a warm-up procedure.

## • Recorder Roles.

Some important recorder roles include:

- Remind the team to stick to non-evaluative listing.
- Keep loud members from dominating the team.
- Encourage quiet members.
- Do not discuss ideas.
- Ask for bizarre ideas.
- Play subdued leadership roles.
- Act as a 'servant' to the creative thinking team.

The opposite of non-evaluative listing encompasses the 'gauntlet.' In many R&D meetings, each idea has to run three gauntlets: (a) people internally filter their own ideas; (b) the recorder's unwillingness to write an idea down; and (c) the R&D team's critical thinking before it takes an idea seriously.

As a result, R&D people don't express most ideas, shoot them down, or never record them. A team finally accepts an idea because one or more people dominate the process or control through status and position. R&D people often prefer their own idea to a better idea, particularly if someone junior proposes it.

Even if you use the gauntlet only 10% of the time, it results in 100% gauntlet. In other words, even if you only evaluate one idea out of ten, people hesitate to share their ideas in R&D meetings and suppress many ideas. See Part 4 of this book for techniques that help during meetings.

---

A HABIT THAT SPOILS R&D CREATIVE THINKING: The **gauntlet**. You need total non-evaluation when listing ideas; you can achieve this with a little practice.

---

### • Buzz Groups.

Use this antidote to the quick fix during R&D meetings. Especially useful when you want many ideas in a meeting without a long discussion.

One person presents a problem. Teams of four to six people turn in their seats to form small 'buzz groups' where they sit. Each buzz group quickly chooses a recorder, who non-evaluatively lists ideas on a writing pad for five minutes. The recorder quickly reads them aloud and gives the list to the presenter of the problem for future use. If the problem-presenter merely says, "Thank you" and does not get into a discussion, the total time takes less than fifteen minutes.

One variation: first ask the buzz groups to non-evaluatively list how-to problem statements. Then one or more of these are chosen by the presenter of the problem for the buzz group to use during idea generation. See Chapter 15 for an excellent use of buzz groups.

A TRUE STORY: In one of my workshops, a plant manager from a Fortune-500 company said he wanted to restructure work groups to eliminate the supervisors and have the teams of workers led by temporary team leaders. Most people favored it except the local union, who insisted everyone get the same pay raise as the team leaders.

I formed three buzz groups with the 18 people present, and while they listed ideas, the plant manager and I privately agreed that at most he might get some trigger-ideas to spark ideas at some later time. To our surprise, the participants generated over 40 ideas, three of which when combined yielded a solution that he thought would satisfy the union, and because of its incentive orientation, he expected to increase productivity while reducing supervisor costs. The plant manager beamed. "Trust the process," we concluded.

## • BRAINWRITING TECHNIQUES

During brainstorming, many people hold back because ideas lack anonymity. By 1970, research had shown that people who sit privately writing their own ideas generated more ideas as a group than a comparable team using brainstorming. An explosion of new brainwriting techniques followed.

People write their own ideas anonymously. Because no one knows the originator of the ideas, status differences do not matter, and people don't hold ideas back as much as during brainstorming. The ideas, while numerous, often lack breadth. However, one can obtain multiple benefits by using both brainstorming and brainwriting techniques in sequence.

A TRUE STORY: One of the participants in a workshop told me during a break that even after I requested non-evaluation and freewheeling idea generation, she still evaluated and held back ideas. She did this even during a brainwriting session where no one would see or judge her ideas. An internal gauntlet.

A HABIT THAT SPOILS R&D CREATIVE THINKING: We continue to carry out **habits** that spoil creative thinking even after we know how negatively they affect us.

## • Idea Gallery Brainwriting.

Write six to ten how-to problem statements at the top of flip chart papers, one problem statement per sheet. Attach the flip chart papers to the wall for idea gallery.

R&D people walk around and write ideas and solutions directly on the papers. The ideas that accumulate on the paper frequently trigger ideas in other people as they wander around.

Try this variation. Hang a paper with a problem statement written at the top outside your office door at work and invite people to write their ideas. Some useful ideas will appear.

### • Idea Card Brainwriting.

Sit quietly for about 30 to 40 minutes, and write one idea per card on 5" x 8" colored index cards with a dark marker so you can read it pinned to the wall. Do not evaluate your ideas. Use the Zen-like non-evaluative listing and automatic writing principles. Occasionally write an absurd, bizarre, exotic idea to trigger other ideas. Now and then exchange cards with other people, relax, and allow someone else's idea to spark new ideas. Record the first idea that comes to mind when looking at each of their cards.

An interesting variation: everyone writes an absurd, bizarre, exotic idea on an index card. Pass the idea card to the person on your right. Write down the first idea that comes to your mind as you read the idea on the card you received. Use that idea as a trigger-idea to spark a better idea.

Another variation: Idea card can help produce an outline for a report or talk. Non-evaluatively list the ideas you would like to include on 3" x 5" index cards, **one** idea per card. Sort your cards in the order you want them to appear in your report or talk. Now prepare your report or talk.

---

A TRUE STORY: I led a four-day problem-solving creative thinking meeting for managers, supervisors, and key professionals from five plant sites of a large Fortune-500 Company to help solve some problems associated with "world class manufacturing." By the end of the meeting, they made committed action plans for five major new approaches and a number of smaller ones. They considered this an outstanding accomplishment.

The following week, a small team made up of people from three plants met to consider the hundreds of ideas that they had overlooked. The process consisted mainly of sorting ideas written on index cards (see Idea Card) and using these ideas as trigger-ideas to spark better ideas. The process worked. What they scheduled as a two-hour meeting

---

lasted until 3 a.m. They identified many new paradigm shifts and developed ten major new proposals.

Idea sorting looking for trigger-ideas after idea card makes a powerful procedure.

### • Clustering Brainwriting.

Corey Ericson, an R&D manager in DuPont, showed me this procedure. Use it spontaneously in a free-writing way.

Write a nucleus-word representing your R&D problem in the center of a piece of paper or a flip chart so all can see. Draw a circle around it. Write rapidly whatever comes to mind in a cluster of words or short phrases around the core nucleus-word. Draw a circle around each word or thought as you write it, and link it to the previous circle. This forms a series of words inside linked circles to form a flow chart of ideas.

Do not think about what you write. Do free intuitive writing. Don't seek connectedness; you want randomness in the early stages. Go off on wild tangents. Shift paradigms. Look for different mind funnels. Seek the bizarre. Stay non-evaluative.

Cluster phrases and words in linked circles around interesting thoughts. Write clusters of words and phrases in linked circles as long as ideas flow freely. When you run out of ideas, stop a while before you add new words or thoughts.

Study the cluster of words and phrases in the linked circles you created. Non-evaluatively list ideas you might want to try. Draw lines between word clusters to form new, synergistic approaches to the problem. Show the clusters of words to someone else to trigger new ideas.

As Corey Ericson says: "Set a quota for a minimum number of ideas. Settling for less than five new ideas accepts the obvious, the quick fix."

To force combinations between unrelated concepts, use **two** or more word-nuclei at opposite corners of the same page. This will force a combination or create a usable metaphor between these concepts as you fill the page.

### • Brainwriting Circles.

A useful variation of clustering.

Write a word or phrase that captures the spirit of the R&D problem in a small circle in the center of a blank sheet of paper on a writing pad or on an easel. Write an entire "how-to" problem statement if you wish.

Write words close to the circle that you associate with the concept within the circle. Use the Zen-like non-evaluative listing and automatic writing principles in a process of progressive free association.

When no more room exists next to the circle, move outward a bit and start writing a new circular layer of words. Continue filling the page with concentric circular layers of words triggered by free association with the original word or phrase, or with any of the words and phrases that you write on the paper.

Stay creative throughout. Use linear and nonlinear creative thinking. Do not evaluate. Go off on tangents. The words you write do not have to connect or make sense.

After you fill the paper, make connections and remote associations by circling and drawing lines between words or phrases that define patterns of new paradigms and ideas that seem useful in themselves or as triggers to new ideas. Use different colored pencils or pens. Make creative associations.

Stay patient. Allow this process to work. It may seems fragile, about to shatter into a useless jumble. Carefully nurture it by not forcing the words and phrases to make sense too soon. Stay with it. This procedure can help your creative thinking in marvelous ways.

## TECHNIQUES USING FORCED-COMBINATIONS AND TRIGGER-IDEAS

Forced-combination helps new ideas appear by mixing the characteristics and properties of two or more objects or thoughts together to spark remote associations. The clock radio illustrates a familiar outcome of forced-combination.

Here's an example of forced-combination between unrelated stimuli. Open a book. Point randomly to any word. List all the properties and characteristics of this word. Combine these properties and your problem to spark new ideas to solve it. Amazingly this quick-fix approach works well.

When you combine very unrelated items and thoughts you produce very original ideas, though these new ideas often prove difficult to turn into practical, workable solutions. Related stimuli and trigger-ideas usually produce less novel ideas easier to use. Make forced-combinations this way:

1, Non-evaluatively list characteristics, associations, reminders, properties of the trigger-idea.

2. Write your problem statement. How to...

3. Non-evaluatively list the ideas from the forced-combination between your problem and the trigger-idea.

**• Analogies as Trigger-Ideas.**

Want exceptional outcomes?

1. Choose a culture, civilization, profession, country, group, organization, animal, or plant.

2. Non-evaluatively list the characteristics of the culture, civilization, profession, group, organization, animal, or plant.

3. Write your problem statement. How to...

4. Non-evaluatively list the ideas from the forced combinations between the characteristics and your problem statement.

**• Metaphors as Trigger-Ideas.**

Metaphors open your mind. Try this.

1. Study an object, situation, or picture.

2. Non-evaluatively list specific characteristics or properties, such as color, shape, texture, odor, feel, sound, taste, composition, etc.

3. Choose one property or characteristic. Non-evaluatively list what that property or characteristic reminds you of. For example, a 'WHITE PAGE IN A BOOK' is...white snow; smooth silk; a flat table top; speckled ash on snow; a lined, plowed field; etc.

4. Non-evaluatively list the characteristics, nuances, impressions, and properties of your metaphor.

5. Force-combinations between the items in '4' above with your problem statement.

6. Select, combine, change, add to, and develop an idea to help generate a quality solution.

7. Improve your idea by non-evaluatively listing, in turn, what you like and can use in your idea; deficiencies that need improving; and ways you can overcome the deficiencies.

8. Use the process again using a new metaphor. Expect good results and you will get them.

**• Metaphors, Poetry, and Creative thinking.**

Use metaphors to write poems about the problem area. For example, suppose you want to improve a BLACK MAGIC MARKER. Examine its characteristics.

- IT HAS BLACK INK like...a squid, a pen, swamp water.

- IT IS BLACK like...midnight, a dark mood, black paint.

- IT HAS A SLEEK SHAPE like...a bullet, rocket, racing car.

- IT HAS A COVER CAP like...a cap on a tooth paste tube.

- IT STANDS UPRIGHT like...a tree, space shuttle, chimney, playpen.

- IT LIES DOWN like...an airplane landing, chopsticks on a table.

- IT IS SHINY like...patent leather, an apple, a mirror.

Combine these metaphors into a poem…

"Ode To A Black Magic Marker In A Creative Thinking Workshop"

"SQUID-LIKE, THE MIDNIGHT BULLET STREAKS THROUGH THE PLAYPEN, EXPLODING INTO TRIGGERS BIZARRE."

In this poem, "playpen" is a metaphor for the workshop, while the bullet (the magic marker) explodes into triggers, the reverse of firing a gun where the trigger leads to the bullet exploding. Write poems in a similar way about your problem statement. This procedure opens up new perspectives and stimulates the creative thinking that leads to quality solutions.

**• Pictures as Trigger-Ideas.**

Choose a picture to trigger new ideas. Non-evaluatively list all visual elements in the picture. Force combinations between specific visual elements and specific elements of the problem. Make remote associations that help devise quality solutions to solve your problem. Use visual triggers from any source and watch your creative thinking soar.

**• Random Words as Trigger-Ideas.**

Combine non-relevant ideas with your problem statement to spark creative remote associations. Put your finger at random on any word below and force combinations between it and your problem statement to generate new paradigms, new problem statements, and new ideas.

• bears, television, cup and saucer, pea pods, mountains, baseball, bamboo, car, Idaho, clock, telephone, tennis ball, sofa, water falls, skiing, dancing, football, Ohio, car, book, house, radio, museum, wine, pencil, watermelon, town, countryside, pen, lamp, cooking, electricity, outer space, pole vault, jokes, sculpture, fishing, candle, rock, Kansas, cows, laugh, joy, fun, toys, dreams, sewing, automobile, religion, laser beam, dice, magic, winter, NYC, meditation, children, rock and roll, astronomy, love, movies, watch, money, friend, school, mountain, laughter, flying, ocean, key, street, store, knife, universe, home, boxing, horses, painting, love, swing.

### • Book pages as Trigger-Ideas.

Use a random word from a book as an idea trigger.

Write down a page number and another number chosen at random. Choose a book. Turn to the page you listed and count words up to the number you listed. Force combinations between that word and your problem statement. Generate new problem statements and new ideas. Simple, flexible, and effective.

### • Quotations as Trigger-Ideas.

Force combinations using your favorite quotations.

1. Non-evaluatively list 10 to 20 famous quotations.

2. Choose a quotation that seems particularly impactful or special.

3. Write your quotation....

4. Non-evaluatively list the quotation's characteristics, properties, impressions, nuances, values, etc.

5. Write your 'how-to problem statement. How to...

6. Force combinations between the characteristics of your quotation and your problem statement. Non-evaluatively list the ideas that come to mind.

7. Select, combine, change, add to, slice, and develop the ideas to solve your problem.

8. Improve your idea. List its characteristics and properties.

(a) Non-evaluatively list what you like and can use in your idea.

(b) Non-evaluatively list deficiencies that need improving in your idea. Convert these into how-to problem statements.

(c) Non-evaluatively list ways to improve your idea by solving the problem statements in '8(b)' above.

9. Repeat the entire process. Allow good results to happen.

### • Idea Grid to Trigger Ideas.

Use idea grid (also called morphological analysis), a systematic search for new mind funnels and paradigms to make sure you did not overlook anything.

Example: "How to improve creative thinking at work?"

1. Draw a grid.

2. Fill in the top and the first side row with categories relating to the main problem.

3. Force combinations between categories to generate new ideas in each box.

**Please fill out the 'Idea Grid' on the next page.**

Please write at least one idea in each category below. If you find some categories not relevant, please change them. Think of other problems in your work which you can solve using Idea Grid.

| | How to Increase Quality of New Ideas in R&D ↓ | How to Increase Quantity of New Ideas in R&D ↓ | How to Help Improve New Ideas, in R&D ↓ | How to Increase Development of New Ideas in R&D ↓ | How to Improve Implementation of New Ideas in R&D ↓ | How to Improve Screening, Selection of New Ideas in R&D ↓ |
|---|---|---|---|---|---|---|
| In R&D Meetings | | | | | | |
| In Yourself | | | | | | |
| In Subor-dinates | | | | | | |
| In Peers | | | | | | |
| In Your R&D Team | | | | | | |
| In R&D | | | | | | |
| In Manu-facturing | | | | | | |
| In R&D Management | | | | | | |

**• Improve Bizarre Trigger-Ideas Game.**

The purpose: To stimulate each R&D person to stretch his or her imagination and express bizarre trigger-ideas beyond previous levels. In addition, to give a creative thinking team the experience of helping to develop an idea no matter how bizarre, so it becomes useful. The game has simple rules.

Each team has four minutes to generate the most bizarre, outrageous, and absurd idea to solve the problem. They pass this idea to another team.

That other team has four minutes to use this idea as a trigger to spark a better idea. In other words, to turn the bears and the pots of honey into helicopters (Chapter 5), or turn the weak seam of the pea pod into the weak seam of a can (Chapter 11). If the team does so, it gets one point. If it does not, the other team gets one point. Expect many unexpected ideas.

**• Weird to Workable Idea.**

The recorder of each creative thinking team divides a large flip chart paper into four equal quadrants with a dark marker.

Each creative thinking team records a very "weird" idea to solve their R&D problem statement in the first quadrant. They make the idea as exotic, absurd, and bizarre as possible. They pass the flip chart paper to another team.

This team uses the weird idea to trigger a "better" idea and write it in the second quadrant of the flip chart paper. They pass the flip chart paper to another team.

This team uses the better idea to trigger a "practical" idea and writes it in the third quadrant of the flip chart paper. They pass the flip chart paper to another team.

This team uses the practical idea to trigger a "workable" idea, writes it in the fourth quadrant of the flip chart paper, and turns the idea into a sensible, practical solution.

In general, the more bizarre and weird the first idea, the more likely the final workable idea captures the unexpected, original, and creative.

---

A FORCED-COMBINATION: Take an imaginary trip to Africa or Venus and bring back something absurd to forcibly combine with the R&D problem. Also, try Asia or Mars as a source of trigger-ideas.

---

**• Free Word Association Imagery.**

Generate relatively few ideas, but the uniqueness of those ideas makes this procedure a gem. Expect new mind funnels and many R&D paradigm shifts. Use five to six people plus a recorder

1. The recorder intuitively selects a dynamic word from the chosen problem statement.

2. R&D people forget about the problem. One person says a one-word free association to the chosen word. Each team member, in turn, says a one-word free association to the word generated by the preceding person. The recorder lists all the words.

3. The recorder intuitively picks one of the words. People close their eyes and spend a few minutes imaging around the chosen word. People describe their images, which the recorder non-evaluatively lists on flip chart paper.

4. People force exotic and bizarre combinations between the images and the problem statement.

5. People use the bizarre combinations to spark practical ideas. The recorder non-evaluatively lists these on flip chart paper. People develop a proposal and a potential sensible, workable solution for the original problem.

The process may be repeated many times! Expect good results.


**• Combining-Ideas Team.**

An excellent idea-generating procedure for an advanced R&D creative thinking group.

1. Form three types of teams consisting of five to six R&D people.

2. Designate one type as the logical team, one type as the creative thinking team, and the third type as the combining-ideas team.

3. The combining-ideas team remains idle while the logical and the creative thinking teams generate ideas to solve a given how-to problem statement using non-evaluating listing and idea card, about 20 to 30 minutes for each procedure.

4. Tell the creative thinking teams to focus on bizarre and wild ideas.

5. Tell the logical teams to concentrate only on logical ideas that make sense.

6. Ask the combining-ideas team to force combinations between the ideas of the creative and logical teams, and develop new creative workable solutions.

A variation of this approach: form only two types of teams, the creative and logical teams, and ask them to force combinations between each other's ideas in mixed buzz groups.

## CREATIVE R&D THINKING TECHNIQUES INVOLVING FUTURE FANTASY

Future fantasy produces the most unexpected, creative, highest quality solutions to complex R&D problems. This approach combines expectations of the future with current reality. Use this only with an experienced creative thinking team.

**• Future Pretend Year.**

A very advanced creative thinking procedure that requires practice.

Write your problem statement. How to...

Fantasize the future when someone solved your problem. List each of the following items in turn on separate flip chart papers without thinking about the possible solution.

1. Non-evaluatively list those people inside and outside your company who gained by the success.

2. Non-evaluatively list those people who lost now that someone solved the problem.

3. Non-evaluatively list the 'Fantasy Resource People' who helped implement the solution. Include...

• Experts from the past and present in your company.

• Experts from other companies or organizations.

• Experts or animals from Sea World, Kennedy Space Center, or Epcot Center.

• Other experts or heroes, historical or mythological, from whom you want help.

• Fantasy helpers: writers, scientists, artists, inventors, thinkers, business people, professors, consultants, etc.

4. For each person or animal listed in '3' above, non-evaluatively list specifically what each uniquely does in the future to help implement the successful solution. Start each item with the name of the person or animal.

5. Use each activity listed in '4' above as a trigger-idea to spark new ideas to solve your problem as it exists today. Force combinations between these trigger-ideas and your problem statement.

6. Improve each idea using the idea-improvement techniques below.

Future pretend year produces a cascade of trigger-ideas, thus:

People => Unique activities => New & Useful ideas.

Allow four to six hours for this very advanced procedure. Don't rush it because it seems non-logical. Logic will play a role, never fear. This procedure: so powerful, an explosion of uniqueness.

## TO IMPROVE CREATIVE OUTPUT AND INNOVATION, IMPROVE YOUR IDEAS

Once you select an idea, you can add to your creative output and innovation by improving it as much as you can. The idea-improving techniques described below helps you accomplish this creatively and may stimulate additional new ideas or provide new perspectives. Creative thinking soars when you creatively improve an idea.

• **Like-Improve Analysis.**

1. Write the idea.

2. Non-evaluatively list the characteristics and properties of the idea.

3. Non-evaluatively list what you like and what you find useful about the idea.

4. Non-Evaluatively list deficiencies that need improving in the idea. List these as how-to problem statements. How to...

5. Non-evaluatively list ways to overcome these deficiencies and improve the idea by responding to the how-to problem statements.

6. Recycle all the above until the idea works.

This procedure works very well even with the most bizarre ideas.

• **Improve Your Idea Creatively.**

Juggle your new idea in your mind until it clicks into place. Do it this way.

• Make it larger, greater, or extend and magnify it.

• Make it smaller, delete it, contract or diminish it.

• Rearrange it, adapt it, transpose it, or substitute for it.

• Consider it from a different point of view.

• Reverse it.

• Distort it.

• Make it stronger, taller, shorter, thicker, lighter, or weaker.

- Draw it.
- Sing it.
- Snort it
- Write a poem about it.
- Split it.
- Massage it.
- Combine and blend ideas.
- Sort them.
- Shuffle them.
- Turn up side down, backwards, opposite.
- Slice them into bits.
- Dance it.
- Make a metaphor of it.
- Imagine how it would work out and change it.
- Imagine what your favorite relative would say and modify it.
- Paint it.
- Eat it.
- Sleep on it.
- Lie on it.
- Roll it up.
- Stomp on it.
- Change its shape.
- Add to it.

Force combinations between your idea and other objects into additional new ideas. Combine remote associations and watch your creative thinking take off.

---

• CHAPTER 13 •

CREATIVE R&D THINKING IN ACTION


TECHNIQUES TO COMBINE IDEAS
INTO CREATIVE TRIGGER-PROPOSALS
(THE 3RD CREATIVE STEP)

---

This chapter deals with Steps 5 to 7 of the eight-step problem-solving sequence that yields high quality solutions.

## FORCED-WITHDRAWAL AND CREATIVE TRIGGER-PROPOSALS

Evaluation uses old information to judge new ideas, so you quickly discard ideas that seem unworkable. This reduces your chances for a high quality solution when selecting ideas. Yet you can convert this normally evaluative (non-creative) step into one that creates new possibilities instead of diminishing them.

One antidote to this difficulty: **forced-withdrawal**. Trick your mind, pretend you own a new company, and combine ideas into a trigger-proposal to solve your problem in the new company, not your current organization. This is Step 5 in the problem-solving sequence. Combine ideas into trigger-proposals, the third key creative step in solving problems.

After you write your trigger-proposals, identify the criteria you will use to evaluate solutions (Step 6). Then force combinations between your trigger-proposal and your real problem, and develop quality solutions for your R&D organization that meet the criteria (Step 7).

Trigger-proposals keep you from committing to an idea too soon, avoiding the quick fix. This unique process forces you to review and combine ideas you might prematurely reject, and thus miss out on a high-quality solution.

This approach keeps you from applying criteria too soon, and allows you to stay creative as you select and combine ideas. Also, it keeps you from using unstated, phantom criteria. Finally, it prevents you from stifling your own ideas with "we tried that before" or "we'll never get them to agree," a typical R&D tactic.

---

## STEP 5. COMBINE IDEAS INTO CREATIVE TRIGGER-PROPOSALS

Suppose you want to develop a quality solution on: "How to stimulate creative thinking in your R&D team."

STEP 5. Prepare an innovative one-page trigger-proposal on how to stimulate creative thinking in your new company (forced withdrawal).

1. Pretend you own a new company.

2. Look over the ideas generated during the idea-generating sessions of step 4. Combine, modify, add to, subtract from, connect, and change them.

3. List ideas you like on 3" x 5" cards, one idea per card. Then non-evaluatively list new ideas about what you might include in a trigger-proposal on 3" x 5" index cards, one idea per card. Let the ideas flow. Let your imagination soar. Stay very creative.

4. Sort the cards. Place the ideas in the sequence you want.

5. Summarize them on one page, your trigger-proposal. Think very creatively.

---

**CHAPTER 14**

**CREATIVE R&D THINKING IN ACTION**


**TECHNIQUES TO IDENTIFY CRITERIA AND**

**DEVELOP QUALITY R&D SOLUTIONS**

---

## STEP 6: TECHNIQUES TO IDENTIFY THE CRITERIA TO CHOOSE HIGH QUALITY SOLUTIONS

Before you can develop a quality solution, you must identify the criteria you will use to choose it. Do this with great care and thoughtful reflection. After all that effort you put into all those techniques, produce a high quality solution. Don't let unstated, phantom R&D criteria spoil your quality outcome. Use the following list to help identify the important criteria:

• Ease of testing and start-up

• The consequences of doing nothing

• Tangible costs:

   materials

   equipment

   space

   personnel

• Intangible costs:

   opinions

   attitudes

   feelings

   aesthetics

• Difficulties of implementation.

• Technical feasibility's.

• Compatible with your work group and the organization.

• Marketability.

• Effect on your overall goals.

 Individuals and teams affected.

• Moral and legal implications.

• New problems caused.

• Consequences of success and failure.

• Timeliness.

• Benefits to you, your work group and the organization.

• Satisfy others who have to agree.

• Your gut feelings.

• Other issues...

Write the criteria you will use to develop proposals and solutions.


## STEP 7. CONVERT YOUR TRIGGER-PROPOSAL INTO A QUALITY SOLUTION FOR YOUR R&D ORGANIZATION.

Develop a fresh, high-quality solution by applying your trigger-proposal for your new company to your R&D organization to solve the real problem at work. Get feedback for improvement from your team and other people.


### • Idea Board

I learned this procedure from William Drath at the Center For Creative Leadership. Write ideas on 5" x 8" index cards and arrange them on a large bulletin board in themes and categories, one idea per card, and one theme or category per column. Make sure you can see the writing from a distance. Attach the cards with pins or for greater ease when arranging the cards, use a magnetic board with magnetized buttons.

An idea board helps sort and categorize the many ideas generated during idea card described in Chapter 12. Use it to display a specific planning process so you can easily see the steps in a plan, and can add new steps in sequence.

Organize the cards in vertical columns on the idea board. Create headings for each column that include all the cards under it. Use different colored cards for different purposes. For example, use yellow cards for general headings and write the ideas on orange cards. Use blue cards to head the column of ideas you have put aside, but not yet rejected, while green cards can head the column of ideas that you think clear winners.

Generate new ideas and new headings as needed. Develop a creative trigger-proposal and a sensible solution that fits your criteria.

## STRUCTURE IS IMPORTANT WHEN GOING AFTER QUALITY SOLUTIONS

By now you might wonder about my emphasis on structure throughout this book. Isn't creative thinking in R&D usually a helter-skelter free-for-all? In reality, it may seem that way, but as my friend, Bob Phillips, pointed out, creative thinking flourishes best for many people within specific structures and special discipline. The intense focus and tuning-out of extraneous stimuli necessary for creative thinking requires Zen-like structure and enforced discipline, not to box the creative person in, but to keep out unwanted distraction and interfering stimuli.

The special conditions vary with the person (see 'motivating catalysts' in Chapter 19). Deliberately set up the structure and stick to it, and relentlessly protect it from breaking down by keeping distractions and interruptions from interfering with the creative thinking process. Assert and say NO to spoilers of creative thinking around you. Don't let the stiflers of creative thinking overwhelm you and prevent you from producing a quality solution.

---

## • CHAPTER 15 •
## CREATIVE R&D THINKING IN ACTION

## READY, SET, R&D ACTION PLANS

---

Achieving quality solutions in R&D involves a probabilities game. Increase the probabilities that you will define the problem and shift paradigms innovatively, generate useful ideas super-abundantly, combine ideas into creative trigger-proposals, identify criteria appropriately, convert trigger-proposals into outstanding quality solutions, and make effective action plans. The payoff in excellence makes this effort worthwhile.

A TRUE STORY: A manager in a Fortune-500 Company asked me to lead a 3-day creative thinking meeting to improve **quality** at work. Thirty-five managers and supervisors in manufacturing and R&D from a large plant attended.

The goals they wanted to achieve challenged me: each person will develop new perspectives and new ideas to help solve their own work-related problems on quality; each person will develop a personal action plan to improve his or her quality at work; each person will use and teach advanced creative thinking techniques at work: each person will learn to define problems creatively, generate ideas abundantly, combine them into creative trigger-proposals, and produce quality solutions.

The meeting consisted of six sessions. **First session**: We carried out techniques to form teams, create a creative atmosphere, and started team building. **Second session**: Selected people gave short talks to teach everyone the elements of quality at work. **Third session**: We carried out techniques to shift paradigms and define the quality problems. **Fourth session**: We carried out techniques to generate ideas abundantly. **Fifth session**: Each team helped solve work related, quality problems of each of its members using the habits and techniques they learned earlier during this event. **Sixth session**: Each person generated trigger-proposals to help solve their own problems with quality at work. They received feedback from their team to improve and upgrade the proposal. They then developed personal action plans.

Excellent outcome. Each person left with personal action plans to improve quality in their work and to help other people improve quality in their work.

## STEP 7. MAKE ACTION PLANS

You accomplish so much more after you make specific action plans containing detailed action steps. Fill out the following alone or with other people:

Action Plan Regarding…

| | Action Step 1. ↓ | Action Step 2. ↓ | Action Step 3. ↓ | Action Step 4. ↓ | Action Step 5. ↓ |
|---|---|---|---|---|---|
| What is to be done? | | | | | |
| Who does It? | | | | | |
| When? | | | | | |
| Where? | | | | | |
| Why? | | | | | |
| What else? | | | | | |

---

# • PART 4 •

## TECHNIQUES FOR CREATIVE R&D MEETINGS

Creative thinking techniques sharpen your competitive edge.

------------------------------------------------------

# • CHAPTER 16 •
## BEEF UP REGULAR R&D MEETINGS

---

Creative thinking in R&D meetings counts. Apply the advanced techniques described in this book to regular meetings of your work group. Use non-evaluative listing and buzz groups (see Chapter 12) at least once in every meeting so everyone looks forward to being creative and solving problems in meetings.

A TRUE STORY: An R&D manager of a large Fortune-500 company told me that he asks his team members to spend 10 to 15 minutes of every meeting using Idea Card to write ideas to solve an important problem. He says it provides him with valuable ideas and it keeps applied creative thinking techniques in everyone's mind.

In addition, apply reversal-dereversal (described in Chapter 11) in your next regular R&D meeting. Ask people to reverse the problem statement "How to *stimulate* creative thinking during our meetings" and non-evaluatively list ideas on "How to *spoil* creative thinking during our meetings." This list often reflects what you all usually do in meetings. Then dereverse each spoiler (as described in Chapter 11): write "How to" in front of each idea and creatively smooth out each sentence into a sensible "how-to" problem statement.

For example, you could dereverse the spoiler, "Have domineering people present" into "How to stay creative with domineering people present" or into "How to keep domineering people out." Reverse another spoiler, "Hold meetings at 4:45 on Friday" into

---

"How to stay creative in a meeting held at 4:45 on Friday" or "How to avoid a meeting called at this time."

You will soon have many problem statements focusing on specific needs of your work group. Form buzz groups and non-evaluatively list solutions to the problem statements that impact the most on your meetings. You all know best what spoils creative thinking during your meetings.

## CREATIVE THINKING SPOILERS IN R&D MEETINGS

The following summarizes what some experts say spoils creative thinking during meetings.

- Members judge ideas prematurely and use quick negative criticism.
- Minimal sharing of ideas occurs.
- Highly vocal people dominate.
- Experts or high-ranking superiors overwhelm.
- People lack training in advanced creative thinking techniques.
- Leaders don't tell people that they want creative outcomes.
- People do not stay interested and involved.
- People focus on achieving the mission, not on new ideas.
- People conceal emotions and inhibit spontaneity and humor.
- People use win-lose methods, such as majority rules.
- People select ideas prematurely, the quick fix.
- People do not solve problems in structured ways.
- People don't know the goals and purposes.
- People use analytical and logical thinking too much.
- The leader encourages ideas most similar to his or her own preconceived notions through verbal and nonverbal feedback.

## CHECK THE SPOILERS ABOVE THAT OCCUR IN YOUR R&D MEETINGS?

## INCREASE CREATIVE THINKING DURING YOUR R&D MEETINGS

Help creative thinking flourish during meetings:

• Use advanced problem-solving creative thinking techniques to:

Define problems creatively.

Generate ideas abundantly.

Select and combine ideas innovatively.

Generate trigger-proposals imaginatively.

Develop workable solutions logically.

Select proposals systematically.

• Postpone evaluation and defer judgment of new ideas.

• Establish a quota for many really different ideas before selection.

• When hearing a new idea, state what you like about an idea first.

• Use effective team interaction techniques, such as:

Make decisions by consensus.

Record on flip charts so all can see.

Allow leadership roles to distribute naturally.

Circulate the agenda before and action plans afterwards.

Rotate the chair among members of the team.

Discuss and review work group interactions frequently;

Review and discuss what spoils creative thinking and productivity.

Study the next page to lead a meeting to help stimulate creative thinking.

See Chapter 26 for self directed team building.


## CHECK THE HELPERS ABOVE THAT OCCUR IN YOUR R&D MEETINGS?

# CHAIR MEETINGS TO HELP CREATIVE THINKING
(Modified from Prince, 1970)

| **Suggested Leader Techniques** | **Reasons** |
|---|---|
| 1. Do not compete with other people to generate ideas. Support and build on the ideas of others. | 1. Leaders tend to favor their own ideas. This discourages other people from contributing. |
| 2. Respond non-evaluatively to new ideas. Create an atmosphere in which people consider all ideas. | 2. Responding in a non-evaluative way encourages everyone to participate. |
| 3. Do not permit anyone to be put on the defensive. Find value in all points of view. Start with what you like about what you heard. | 3. This approach encourages everyone to contribute and help new ideas. |
| 4. Get people to talk about the positives of an idea before the negatives. Do not kill an idea; just put it aside. | 4. This approach encourages everyone to contribute and help new ideas. |
| 5. Keep your energy level high. | 5. Your interest and alertness helps. |
| 6. Use every member of your work group. Talk to domineering people privately. Help quiet persons. | 6. Everyone has unique mind funnels, valuable ideas, and information that contribute to the quality of the outcomes. |
| 7. Tape meetings and ask persons with poor behavior to listen to the tape. | 7. This helps them to change their behavior. |
| 8. Rotate the chair of the meeting. | 8. Being a follower and leader leads to commitment and participation. |
| 9. Do not damage egos or self-esteem. | 9. This encourages everyone to share and leads to greater levels of participation. |
| 10. Defer judgment during idea generation and avoid early commitment to an idea. | 10. The leader has great power to sway members. This does not always result in choosing and developing the best idea. |

---

# • CHAPTER 17 •
## TECHNIQUES FOR PERMANENT R&D
## 'CREATIVE THINKING TEAMS'

Permanent creative thinking teams consist of R&D and other people in your organization trained in diverse creative thinking techniques and who can generate numerous problem statements and many ideas to help solve other people's problems. Do it this way.

First: chose a temporary head of the team, an experienced creativity member, to make the arrangements and act as recorder during the creative session. In addition, this person will coach the R&D problem-presenter in the roles he or she must play in order for the process to work and help clarify the problem presented to the team.

Second: chose the five to eight members of the team. They provide the different perspectives and fuel the creative energy to generate endless how-to problem statements and new ideas. Their only reward: the inner fun of creative thinking and helping others to solve problems.

Finally: the problem-presenter must present the problem clearly to the team, so spend time to make sure this occurs. As an expert able to present information, the problem-presenter should stay non-evaluative and temporarily suspend judgment throughout to help the process work. A unique role of the problem-presenter: choose the problem statements for the team to tackle; no one else has this choice.

---

A TRUE STORY: I presented a one-day creative thinking workshop for a Director of Information Systems of a Fortune-500 company and 33 of his managers and computer analysts. I formed six teams who worked on a problem presented by their Director.

They helped define the problem by suggesting over 50 problem statements. He chose six of these, one for each team. They generated hundreds of ideas using non-evaluative listing, improve bizarre ideas game, idea gallery, and idea card.

They had a good time and the problem seemed on the way toward solution. Later, he asked me the best way to sort and select ideas.

---

## CREATIVE THINKING TEAM SESSIONS

The sequence of creativity techniques and the time allotted for a session will vary with the importance of the problem. Sometimes a 60 minute buzz group session does the job. Sometimes you will need an entire day, or more, using many techniques for defining problems and generating ideas. A sequence of creative thinking techniques I find useful includes:

(1) The problem-presenter presents the problem.

(2) The recorder non-evaluative lists the how-to problem statements suggested by the team. (Set a quota for 10 to 20 how-to statements.)

(3) The problem-presenter chooses the problem statement on which the team will focus.

(4) The recorder non-evaluatively lists the ideas suggested by the team. (Set a quota of 30 to 50 ideas.)

(5) The recorder non-evaluatively lists 10 to 15 bizarre trigger-ideas suggested by the team.

(6) The team combines and improves the most bizarre ideas.

(7) Each member of the team then sits quietly alone and writes ideas on 5x8 index cards, one idea per card (Idea Card).

(8) Give all the papers and cards to the problem-presenter.

(9) The team then applies 'like-improve analysis' to how the team functioned. The recorder non-evaluatively lists what each person liked and what each wants improved to make the team even more effective.

During the session, the problem presenter presents a low profile, does not act defensively, does put any idea down ("we thought of that one" or "we tried that one"), and thanks everyone at the end for the excellent ideas generated. If asked if any of the ideas seem useful, answer an enthusiastic "yes," an important reward for the team.

---

A TRUE STORY: Shortly after I presented a creative thinking workshop to an R&D unit of a Fortune-500 company, I received a call from one of the managers to tell me about a participant.

"He's an excellent engineer but never listened to his subordinate's ideas. After your workshop, he started conducting meetings with them using the methods he learned in your workshop. Yesterday, he and one of his subordinates showed me an idea the

---

> subordinate had suggested and it's terrific. Very simple and it should be easy to implement. We will see people in marketing about it next week." He pointed out this innovative product could pay for all the workshops I had presented in his division.

## WRITE UP YOUR PROBLEM AS PROBLEM-PRESENTER

1. Does your problem have to do with a situation, individuals, a work group, or a thing?

2. Outline your problem. Make sure you have responsibility for the problem and can implement solutions.

Write a problem statement: How to…

3. I would like this problem resolved because...

4. List some possible indicators of success.

5. List the resources available to help resolve the problem.

6. List the obstacles you have to overcome.

7. Any deadlines?

8. Other issues?

9. In two to four sentences, write a broad-brush over view of your problem.

10. List what you might lose if the problem continues.

11. List what you will gain if you solve the problem.

12. List the approaches and solutions you have already tried and why each has failed.

13. List the benefits of the status quo and the advantages of doing nothing.

14. Non-evaluatively list dozens of how-to problem statements and chose 2 to 4.

15. Summarize your problem in one problem statement:

 How to...

16. Whose problem?

17. What kind of problem?

 Marketing:

 Manufacturing:

 Technological:

 People:

 Financial:

Other:

18. How big?

Funds:

People:

Time:

Other Resources:

19. Your gut feelings about the problem?

20. Intangible Issues?

21. Other issues?

List the criteria to select your problem statement to give to the team.

---

A TRUE STORY: People in a large Fortune-500 asked me to present a three-hour workshop for 100 managers on "market-driven quality." I formed teams of six people each, and even in this short time they used non-evaluative listing and idea card to harvest hundreds of ideas on 'how to achieve market-driven quality at the plant site.' These ideas were turned over to the delighted site director, who asked for, and got, volunteers to help sort the ideas for feasibility. The outcome astonished everyone, showing that low-level techniques can achieve happy results even when untrained people use them.

---

**• CHAPTER 18 •**

**TECHNIQUES FOR AN R&D CREATIVITY & INNOVATION MEETING TO SOLVE IMPORTANT PROBLEMS OF YOUR ORGANIZATION**

---

Whether you use advanced creative thinking techniques in R&D team meetings or when you work alone, you achieve high-quality solutions to important problems. These techniques work even better in a '**Creativity & Innovation Meeting**.'

I have led such meetings for large and small companies to solve diverse problems, including:

• raising quality at work;

• identifying new products;

• improving chemical yield during a complex manufacturing process;

• reducing waste at work;

• applying world class manufacturing principles to a product;

• lowering costs and increasing effectiveness of environmental cleanup for a chemical company;

• developing a new technology for manufacturing a specific product;

• handling manufacturing waste for an automobile parts firm.

I consider Creativity & Innovation Meetings the most effective way to solve important R&D problems at work. Creative thinking facilitates the recombination of old ideas and elements in your mind into new and useful solutions. Advanced creative thinking techniques expand this process; this helps you combine more diverse bits from your environment and in your mind into incredible new and useful ideas.

I also consider this meeting the best way to teach advanced creativity techniques: the participants are motivated to learn the techniques to apply to the problem.

**The Sequence Within Each R&D Creativity & Innovation Meeting**

Although each Creativity & Innovation Meeting differs from the others, the same basic sequence appears in all. The following illustrates a typical flow for a four day meeting:

---

-- First session: Introductions; review the goals and agenda; initiate team building within each team; use techniques to create a creative atmosphere; learn that trigger-ideas spark creative ideas during linear and nonlinear creative thinking.

-- Second session: The meeting organizer explains the problem and some participants give short talks in their areas of expertise, when necessary. After that, individuals, and then each team, define the problem using advanced techniques.

-- Third session: Each team produces ideas using many techniques, after which each individual generates ideas sitting quietly alone. Participants usually produce about eight hundred ideas that we display in the meeting room.

-- (Fourth session: An afternoon of free incubation time plus some individual fun work to enhance creative thinking.)

-- Fifth session: Each person combines some of the ideas displayed in the room into an innovative one-page proposal for fresh approaches to solve the problem. The teams then identify the criteria for an effective, high quality solution. Each participant shares their one-page proposal with his or her team and receives feedback for improvement based on the criteria the team identified. Each person then revises the one-page proposal and gives it to management.

-- (Sixth session: An afternoon of free incubation time to enhance creative thinking.)

-- Seventh session: Each team combines the proposals of its members and develops a quality solution to solve the problem.

-- Last session: Each team presents its quality solution to the other participants and receives ideas for improvement from everyone, an exciting, constructive time. Then the participants commit to action plans that implement the best solutions.

## Outcomes of an R&D Creativity & Innovation Meeting

The outcomes of a R&D Creativity & Innovation Meeting fascinate and gratify. Participants say they enjoy the time well spent. They write positive evaluations ("I wish I had learned this stuff 15 years ago"). They learn techniques to achieve quality solutions and most people say that they will use them back on the job. (Sometimes, just before the end, people make plans to spread creative thinking techniques throughout the organization.)

Management receives hundreds of ideas to solve the problem, including some high-quality gems. In addition, each participant hands in a one-page proposal, many containing fresh and unexpected approaches. And each team produces a unique, high-quality solution. Sometimes participants combine solutions from different teams and develop them further. After one spectacular meeting, the senior R&D person in charge told me that fifteen to twenty patentable ideas emerged.

---

A SUCCESS STORY: People in a Fortune-500 company asked me to lead a four-day Creativity & Innovation Meeting to solve their "world class manufacturing" problems. Managers, supervisors, and key professionals in R&D and manufacturing from five plants attended. They wanted to shift paradigms, gain new perspectives, produce new ideas to improve manufacturing practices, and capture long range, world wide competitive leadership.

They sought to accomplish their goals using modern creative thinking techniques. They also wanted to attain synergy among the participants, share what's happening at different locations, and promote networking and teamwork between each plant site. And they wanted specific action plans and commitments from the participants at the end of the meeting, a tall order.

We met in Washington, DC for four days and applied advanced techniques to problems they unearthed early in the meeting. The participants spent two free afternoons in the NASA Space Museum and parts of the Smithsonian Institution. By the end of the four days, they shifted many paradigms and produced an incredible number of outstanding ideas and quality solutions that exceeded the most optimistic predictions. They easily made committed action plans.

---

A SUCCESS STORY: One executive of a company wrote me "to declare the creative thinking sessions a roaring success. Besides the excellent ideas generated, and there are a handful of outstanding ones, it served the purpose of bringing our management team closer together... Thanks for an outstanding job."

---

### Participants

Whom you include in a Creativity & Innovation Meeting depends on the problem and other goals. For example:

• Forty-two people attended a meeting arranged by a vice-president of marketing to develop new products, including the chairman of the corporation, its president, the vice-presidents and managers of sales, marketing, manufacturing, engineering, and finance, and the directors of personnel and quality. The other people included key professionals in sales, marketing, customer relations, industrial design, manufacturing, engineering, and finance, a wide range of people chosen intentionally to solve the problem.

• In a meeting designed to attack environmental-cleanup problems, the organization included managers and key professionals from R&D and maintenance, as well as outside consultants, university professors, people from government, and experts from other companies. Very diverse participants.

• A consulting engineering firm and its client, a chemical company, worked together in a meeting that contained vice-presidents, managers, supervisors, and key professionals from engineering, R&D, marketing, manufacturing, and finance.

• One organization included customers with whom they worked to solve mutual problems in a meeting that comprised managers, supervisors, department heads, and key professionals from marketing, R&D, and manufacturing.

In short, Creativity & Innovation Meetings may include executives, managers, supervisors, department heads, key professionals, and other people from manufacturing, R&D, human resources, finance, maintenance, and marketing. The mix and number of each level and function of the participants depends on what the organization wants to accomplish.

### Designing the R&D Creativity & Innovation Meeting

I custom design every Creativity & Innovation Meeting. To ensure success (the highest quality solutions), I build the following into every design:

Working In Teams and Alone

Many people believe that high-level creative thinking demands that people work alone. Still, teams accomplish truly unexpected, outstanding quality solutions, especially when the team fosters the individual creative thinking of each team member.

I use techniques for teams and individuals in each major step of the problem-solving process. The use of powerful techniques for teams together with equally powerful techniques for participants working alone results in an exciting mix of team interactions fueled with the creative thinking of each individual. Out of this mixture comes success.

## The Importance of Incubation Time

Incubation time, a stage in the innovation process, represents an important design-factor. The stages in the creative process include (see Chapter 5): Preparation; Concentration; Incubation; Illumination; and Implementation. I apply this theory when I design Creativity & Innovation Meetings:

--We choose expert participants who come to the meeting with prepared minds (the Preparation Stage).

-- People focus on the problem during the meeting (the Concentration Stage).

-- People pay attention to other things during free afternoons, evenings, and overnight (the Incubation Stage).

-- Every morning-after provides an opportunity for new insights (the Illumination Stage).

-- Most people leave the creativity & innovation meeting committed to action plans (the beginning of the Implementation Stage).

The Incubation Stage partly explains why the most effective meetings last at least four days. This length helps the incubation processes in the mind so participants make unexpected connections and blockbuster ideas appear as the subconscious mind does its work. Three-day meetings also succeed, though not quite as well.

I insert incubation time into the design of each meeting and often plan specific adventures. For example, during free afternoons in Washington, DC, participants visited the Smithsonian Air And Space Museum (and the Museum of Natural History or an Art Museum) and wrote metaphors and poems about the problem while they incubated the problem there.

And during a six-day meeting in Orlando, FL, participants spent many afternoons and evenings at Epcot Center, Sea World, and the Kennedy Space Center. People used the idea triggers and metaphors they found on these outings to spark creative thought in the meeting.

## Separate The Front and Back End

I design the front and back end of the Creativity & Innovation Meeting as consecutive, non-overlapping processes. The front end focuses on creative thinking, while the back end on logical, evaluative thinking.

The one or two day problem-solving meetings that an organization often carries out alone do not last long enough to yield truly high quality solutions. The front end of the meeting becomes shallow and rushed, so participants generate common and ordinary ideas, instead of fresh, unexpected ones. The solutions usually consist of ideas some participants carried with them when they entered the meeting. Shallowness also exists during the back end, as participants overlook many excellent possibilities by not using forced-withdrawal and trigger-proposals, stunting the development of high quality solutions. Such home grown meetings often wind up with solutions previously thought of by management or other people, instead of creating new ones.

I build in enough time for the front and the back end so the participants create unexpected high-quality solutions, not a quick fix. That's why I prefer a four-day meeting: creative thinking sparkles and the participants generate more creative and unexpected ideas; each team shifts more paradigms; and management receives higher-quality solutions that lead to success.

**Advance Planning Session**

About a month before each Creativity & Innovation Meeting, I attend a planning session with people in the organization to discuss the problem(s) and to agree on goals. This takes place on the site of the organization because I want to meet many of the participants while I listen for the organization's needs.

After a break for reflection, I outline the kind of meeting I think the organization needs to achieve the high caliber solution they desire. I point out that if they want fresh and unexpected solutions, they need a Creativity & Innovation Meeting. If they want solutions already in someone's mind, they can take a poll and make a list. Achieving high quality solutions requires the time to use many advanced techniques to create the unexpected.

I suggest that the meeting takes place away from the plant site, preferably out of town. I don't want the participants disturbed or tempted to visit the office during breaks. I advise them that in my experience, four-day meetings produce more effective outcomes than shorter meetings. I talk about incubation time, the slow processes of creative thinking, and avoiding the quick fix. I emphasize success.

After I return home, I develop a minute-by-minute design and send it to people in the organization for their comments.

They assign people to the small groups (or creative thinking teams) before the meeting. I advise them to put people at the same level in the same team (no executives with supervisors; no bosses with subordinates), and to mix the people in each team according to functions, locations, departments, and talents. I suggest they put five to six people on each team, though a seven-person team can work well together. Still, no one can predict how a team will react; that's one reason to form many teams. If one team performs poorly, others will perform well. Because I insert team building elements into the design almost all teams do perform well.

I ask each team member to read different parts of the same creativity book before the meeting, so knowledge of creative thinking that leads to quality solutions diffuses within each team. I also ask each person to bring preconceived ideas for quality solutions along with a willingness to replace these pet notions with fresh ideas for success.

Finally, after all this, I feel ready to lead this creativity meeting.

## Team Building During An R&D Creativity & Innovation Meeting
### • Forming the small groups (creative thinking teams).

I like to use six-person teams. This provides one recorder and five other group members. If one group member leaves the room, the team members can still participate effectively. If the group has only five members to begin with, then the loss of one member leaves only one recorder and three members, usually too small for effective work.

I first form groups of six and then add additional people, one at a time, to each team to make groups of seven. More than seven members creates difficulties: competition for air time and span of attention, although eight-person teams have occasionally worked well.

I aim for a diversity of different types of people on a team. In addition, I keep status and job levels equal. If unavoidable, I ask high-status persons and experts not to doom the creative climate. I encourage the low-status team members to participate.

### • First Session: Introductions within the small groups within the Creativity & Innovation Meeting.

Everyone wears a name tag throughout the meeting, even me. The first interaction within each team consists of 'introductions,' even if everyone knows each other. I ask each member of the team to share something about themselves that no one else in the group knows about. For example: share one creative idea or innovation of yours in your organization; share a funny experience; share something you are proud of; etc. I want this first interaction of the team to remain low key and allow each person to contribute something unique.

**• Warm up using riddles and puzzles.**

After introductions, I ask each person to solve a puzzle alone, and then working with the team. I discuss the answers in terms of habits that affect the creative atmosphere in the meeting. Examples include "IX" and "half-of-8" in Chapter 6; "9-dots" in Chapter 7.

**• Another first-session warm up.**

Soon after the first session starts, I carry out a warm up exercise that teaches or reminds people about non-evaluative listing (see Chapter 12). I ask each team to list items (for example, list ways to spoil creativity at work; list ways to spoil meetings; list ways to spoil teamwork; list what you want to accomplish here, and then I describe non-evaluative listing, the gauntlet that new ideas run, and the recorder's roles. I emphasize the necessity of a creative atmosphere during the meeting and that everyone has 100% responsibility for the creative efforts of others and for themselves (see Chapter 20).

• Starting each day.

I start each new day of the meeting with a low key interaction within each team. For example, I ask them to share new ideas they thought of overnight; funny experiences; jokes; dreams that triggered new ideas; etc. I want people to connect with each other each morning with a low key task that all can share. Sometimes, I remove the captions from New Yorker cartoons and hand several cartoons to each person; I ask them to write captions for the cartoons overnight. Laughter is a wonderful way to start the day's interactions.

**TECHNIQUES TO USE IN AN R&D CREATIVITY & INNOVATION MEETING**

I list here the steps and techniques of creative thinking described in this book in a sequence to achieve the highest quality solutions to the problem at hand. Your creative thinking will soar as you use them on a regular basis. Learn them by using them on important problems, especially recurring problems that lack clear focus. The following steps create focus and clarity:

## STEP 1. Define Problems Creatively, the 1st Key Creative Step (Chapter 11).

- List Dozens of Problem statements
- Analogies And Metaphors
- The Problem's Essence
- Analogies And Metaphors Using The Problem's Essence
- Like-Improve Analysis
- Reversal-Dereversal
- Reverse Assumptions
- Guided Fresh Eye Approach
- Word Substitution
- Why, Who, What, Where, When, Why
- Needs, Obstacles, And Constraints
- Weaknesses Of Quick-Fix Solutions

## STEP 2. Identify The Criteria To Select Your Problem Statement (Chapter 11).

## STEP 3. Choose The Final Problem Statement(s).

## STEP 4. Generate Ideas Abundantly, the 2nd Key Creative Step (Chapter 12).

- Brainstorming
- Non-Evaluative Listing
- Buzz Group
- Idea Gallery
- Idea Card
- Clustering
- Brainwriting Circle
- Forced Combinations and Trigger-Ideas
- Analogies as trigger-ideas
- Metaphors as trigger-ideas
- Pictures as trigger-ideas
- Random word as trigger-ideas
- Quotations as trigger-ideas
- Idea Grid

- Improve Bizarre Trigger-Ideas Game
- Weird To The Workable Idea
- Free Word Association Imagery
- Combining-Ideas Team
- Future Fantasy
- Future Pretend Year
- Idea Improvement
- Like-Improve Analysis
- Improve Your Idea Creatively
- Idea Improvement Checklist

## STEP 5. Combine Ideas Into Trigger-Proposals Innovatively, the 3rd key creative step (Chapter 13).
- Forced Withdrawal
- Trigger-Proposals

## STEP 6. Identify The Criteria To Choose Ideas (Chapter 14).

## STEP 7. Develop Sensible Workable Solutions (Chapter 14).
- Return To Reality
- Idea Board

## STEP 8. Make Action Plans (Chapter 15).

## A SPECIAL SEQUENCE TO GENERATE IDEAS IN A R&D CREATIVITY MEETING

One idea-generating sequence I like to use during a Creativity & Innovation Meeting consists of non-evaluative listing (brainstorming) => improving bizarre trigger-ideas game => from the weird to the workable idea => idea gallery => free word association imagery => idea card.

This sequence starts with the walls covered with flip chart paper on which we have non-evaluatively listed over 150 "How-to" problem statements using techniques described in Chapter 12.

I ask the people in the meeting to walk around the room and check five to ten problem statements which they would like to solve, a straw vote to identify the problem statements that most people deem important.

I then ask each team to choose the three problem statements that they want to solve. Note, we do not kill any problem statements, we merely leave them behind as we move forward.

### • Non-Evaluative Listing.

I like to start with non-evaluative listing (brainstorming) because everyone records their pet ideas and they stop worrying that they will lose them. Also, it flushes and cleans out the mind of obvious solutions and makes room for newer ideas, so the more advanced techniques work even better. Consider brainstorming and non-evaluative listing a nice warm-up procedure, at best.

I ask each team to choose a recorder and non-evaluatively list (brainstorm) ideas on flip chart paper to solve one of the how-to problem statements. I occasionally ask for more bizarre ideas. Much laughter.

I ask each creative thinking team to hang the flip chart papers containing their ideas on the wall for future use. They quickly notice that the how-to problem statement channels the ideas generated. Thus, if you want to solve the right problem, choose the right problem statement.

### • Improve Bizarre Trigger-Ideas Game.

I now suggest that they do not understand what I mean by a bizarre idea, and ask them to play the improve bizarre trigger-ideas game.

I want to get each person to stretch their imagination and express bizarre ideas beyond previous levels. In addition, it shows the team how to develop an idea no matter how bizarre, and strengthen it so it becomes useful. Paradigms shift constantly.

The game has simple rules. Each creative thinking team has four minutes to generate the most bizarre idea to solve the problem. They pass the idea on paper to another team.

That other team has four minutes to use this idea as a trigger to spark a better idea. If it does so, it gets one point. If it does not, the other team gets one point.

I repeat this game two or three times. It amazes me how bizarre the ideas can get and how ingeniously people use them to trigger useful ideas. I sometimes pass out folded blank flip chart papers as the bizarre idea. Even a folded blank sheet sparks new ideas.

This idea-generating procedure triggers a turning point. People discover that openness to bizarre ideas through non-evaluation can pump up creative thinking and precipitate paradigm shifts.

### • Weird to Workable Idea.

I now ask the recorder of each team to divide a large flip chart paper into four quadrants with a marker. Each creative thinking team writes a very "weird" idea to solve their problem statement in the first quadrant. They generate an exotic, absurd, and bizarre idea. They pass the flip chart paper to another team.

This team uses the weird idea to trigger a "better" idea, and writes it in the second quadrant of the flip chart paper. They pass the flip chart paper to another team.

This team uses the better idea to trigger a "practical" idea, and writes it in the third quadrant. They pass the flip chart paper to another team.

This team uses the practical idea to trigger a "workable" idea, and writes it in the fourth quadrant of the flip chart paper. They turn the idea into a sensible, practical solution. In general, the more bizarre and weird the first idea, the more likely the final workable idea shifts a paradigm and creates a different, original, unexpected outcome.

### • Idea Gallery.

Earlier, I noted the six to ten problem statements that people checked the most during the straw vote. I write these problem statements at the top of flip chart paper, one problem statement per sheet and attach them to the wall for idea gallery.

People walk around and write ideas and solutions directly on the papers. The ideas that accumulate on the paper frequently triggers ideas in other people as they wander around. Such movement often helps creative thinking.

### • Free Word Association Imagery.

You create very few, but highly unique and worthwhile ideas. Use five to seven people plus a recorder for best results.

1. The recorder intuitively selects a dynamic word from the chosen problem statement.

2. People forget about the problem. One member says a one-word free association to the chosen word. Each member of the team in turn gives a one-word free association to the preceding word generated.

3. The recorder intuitively picks one of the words. People close their eyes and spend a few minutes forming an image around the chosen word. People describe their images, which the recorder non-evaluatively lists on flip chart paper.

4. People force exotic and bizarre combinations between any of the images and the chosen problem statement, the more impractical, absurd, and utterly weird, the better.

5. People use the bizarre image combinations to trigger practical ideas which the recorder lists on flip chart paper.

6. The team develops the list into a quality solution.

The process may be repeated many times! Expect extraordinary outcomes and many paradigm shifts.

## • Idea Card.

People sit quietly for about 30 to 40 minutes and privately write one idea per card on 5" x 8" colored index cards with a dark marker. I ask them to write non-evaluatively. I suggest they use the Zen-like automatic writing principles (see Chapter 8).

Occasionally I encourage them to write an absurd, bizarre, exotic idea to trigger other ideas. After a while I suggest they exchange cards in order to relax and allow someone else's idea to spark new ideas. I encourage them to write down the first idea that comes to mind when looking at each new card. I call a five-minute break after 20 minutes.

As another variation, I ask them to write an absurd, bizarre, exotic idea on an index card and pass the idea card to the person on their right. I suggest they use that idea as a trigger-idea to spark a better idea and write down the first idea that comes to mind as they read the idea on the card from the other person.

Idea card gives each person a chance to sit quietly and thoughtfully reflect on the problem, mull over the new mind funnels and paradigms they have discovered, and to privately generate new ideas to solve the problem. Many new ideas emerge.

After idea card, we sort the cards, place them on tables or the floor, or pin them to a wall so the participants can see them. See idea board in Chapter 12 for an excellent sorting procedure.

**• Summary.**

This sequence of advanced idea-generating techniques teaches people, in turn,

- to stay non-evaluative and get rid of pet ideas and open their minds to fresh ideas (non-evaluative listing);
- to use bizarre ideas to trigger better ideas (improve bizarre ideas game and weird idea to workable idea);
- to generate bizarre ideas and force combinations between them and the problem (free word association imagery);
- to quietly record ideas alone (idea gallery and idea card).
- And best of all, as all this learning takes place, many paradigms shift, and people create hundreds of ideas to produce high-quality solutions.

---

# • PART 5 •

## TECHNIQUES FOR R&D PEOPLE WHO WORK ALONE

------------------------------------------------------

### • CHAPTER 19 •
### THINKING CREATIVELY ALONE

---

If you have drifted into routine ways when working alone, change your habits and use techniques to help you improve your creative thinking and ensure success. The following may also indicate a need to boost the way you think:

• You always think the same way...or think very little.

• You keep doing the same things in the same ways.

• You lack interest in what you do.

• You don't take risks in your work.

• The R&D work does not challenge you enough.

• You ask few challenging questions.

• You have resistance to trying out something new.

It may help if you consider some of the key elements in creative thinking techniques.

## SOME BASIC ELEMENTS IN EFFECTIVE CREATIVE THINKING TECHNIQUES

Advanced creative thinking techniques contain basic elements and underlying principles. Once you appreciate their fine points, you can use them more effectively. You can even design your own techniques for when you work alone by combining elements from different techniques to fit special needs. Some key elements include:

### • Avoid premature criteria.

Knowing the criteria for a quality solution spoils creative thinking. Criteria box you in, and you waste time worrying whether each idea and new perspective meets the criteria. Instead, dump criteria. Distort, ignore, and forget stated and unstated phantom criteria.

Phantom criteria include criteria you made up or carry unawarely in your mind. You think they apply, but they don't. No one told you to use them. Or reverse the criteria as described in Chapter 12.

### • Forced-withdrawal.

Change the setting of your perspective. Create and combine ideas within a different context than the real problem. For example, pretend you work for a different company. In this way, you avoid getting bogged down in stifling thoughts and habits. Forced-withdrawal helps you escape the constraints of the real problem and provides you a clearing within which to stay creative. See Chapters 7, 13, and 14 for details.

### • Trigger statements.

Trigger-statements include problem statements that will not work, but when properly used as triggers sparks other problem statements that shift paradigms and change perspectives. They help you avoid time worn paradigms and lead you down new mental paths.

### • Trigger ideas.

Trigger-ideas, ideas that do not contribute to a quality solution, when properly used, trigger other ideas that do. Trigger-ideas play a key role in many advanced techniques.

### • Triggered free association.

All ideas trigger new ideas in unstructured ways. Stay open to this possibility when pursuing a quality solution. Indeed, words can spark new ideas.

### • Forced combinations.

The creation of unexpected and useful ideas depends on you combining ideas, objects, thoughts, and impressions with your problem statement, one cornerstone of creative thinking.

You may combine your problem with thoughts related or unrelated to the problem. Combinations with unrelated items produce very creative results hard to apply to the problem. Combinations with related items yield less novel outcomes often easier to apply to the problem. Take your choice, but why use advanced creative thinking techniques to produce prosaic results (see Chapters 8, 11, and 12).

**• Improve the idea (Like-improve analysis).**
List what you like about the idea so you won't change that, while you list what needs improvement. Use this approach in a variety of techniques: defining problems; improving ideas; analyzing quality solutions.

**• Focused idea improvement.**
Improve the ideas you create. One approach expands like-improve analysis:
Step 1. List the characteristics and properties of the idea.
Step 2. List what's useful and what you like about the idea.
Step 3. List deficiencies in the idea that need improving.
Step 4. List ways to overcome deficiencies and improve your idea.
Step 5. Recycle the above until the idea shines.

**• Combine ideas into trigger-proposals first.**
Sort and combine ideas into a trigger-proposal and then transform it into a quality solution. A trigger-proposal is based on forced-withdrawal and needs transformation into a quality workable proposal. This new and important approach prevents the premature use of criteria that can ruin a potentially high-quality solution when choosing ideas. See Chapter 13 for details.

**• Future fantasy.**
The unhurried use of fantasies about the future often produces the most creative, highest-quality solutions to a complex problem. Combine expectations of the future with current reality, potentially a most successful procedure (described in Chapter 12).

## ADAPT CREATIVE THINKING TECHNIQUES TO YOUR OWN NEEDS

Adapt the techniques described in this book, especially those described in Chapters 11-14, to use when you work alone. Stay creative. What works counts. Avoid philosophical discussions about the nature of creativity. Allow no boundary to limit how much you change a procedure to fit what you want.

- Shift paradigms and define problems creatively to produce quality solutions.

- Avoid the internal gauntlet and other habits that distort the creative atmosphere in your mind.

- Use forced-withdrawal with metaphors and analogies to let your subconscious mind creep in.

- Stay patient and relentless.

- Use forced-combinations and free association.

- Apply the Zen-like automatic writing and non-evaluative listing principles to brainwriting.

- Especially look at clustering and brainwriting circles in chapter 12 and forced withdrawal, trigger-proposals, and idea board in Chapter 13.

## PRODUCTIVE SEQUENCE OF IDEA-GENERATING TECHNIQUES WHILE WORKING ALONE.

Use this proven sequence of idea-generating techniques (step 4; adapted from techniques described in Chapter 12) when working alone.

The ideas you generate using this sequence will astonish you with their unexpectedness, freshness, and usefulness. Each technique in the sequence builds on the previous one and reaches into higher levels of creative thinking. Expect amazement and delight at the outcome, an explosion of unique ideas.

Start by generating numerous "How to" problem statements (not less than 50; 150 is better) using the techniques described in Chapter 11. Record these on regular paper if you wish, but I prefer to move around writing on large flip chart paper on an easel or the wall. This movement seems to help my creative thinking. Or you may record them on index cards, one idea per card to facilitate sorting. See Chapter 12 for 'idea card.'

Check the six to ten problem statements that you find most interesting and impacting. From these, choose three quite different problem statements to tackle. Now start the techniques (adapted from Chapter 12) to generate ideas.

### • Non-Evaluative Listing.

Non-evaluatively list ideas for one of the problem statements you chose. After five minutes, include some silly bizarre ideas. When you run out of ideas for one problem statement, start the non-evaluative listing process on another. Work on at least three quite different problem statements.

### • Improve Bizarre Trigger-Ideas.

Non-evaluatively list at least fifteen bizarre and absurd ideas to solve a problem statement. Stay really bizarre. No one watches you when you work alone.

Then combine the most bizarre ideas and use the outcome as a trigger to spark a better idea. In other words, turn the bears and the pots of honey into helicopters (Chapter 5) or the weak seam of the pea pod into the weak seam to open a can (Chapter 11). Use the bizarre ideas ingeniously to trigger useful ideas.

Finally, fold blank flip chart papers into different shapes and use them unfolded as the bizarre trigger-ideas to spark better ideas. Even the blank paper sparks an active imagination at this stage, and all sorts of interesting ideas emerge. Also use driftwood or strangely shaped rocks as triggers to spark ideas.

### • Weird To The Workable Idea.

Fold a paper sheet into four equal quadrants. Then think weirdly: create a very weird idea. Write it in the first quadrant. Make the idea as exotic and bizarre as possible.

Use the weird idea to trigger a better idea and write it in the second quadrant.

Use the better idea to trigger a practical idea and write it in the third quadrant.

Finally, use the practical idea to trigger a workable idea and write it in the fourth quadrant. Turn this idea into a sensible, practical solution.

Repeat as needed. The more bizarre and weird the first idea, the more likely you will produce an unexpected and unusual workable idea at the end, a paradigm shift.

### • Idea Gallery.

Write six to eight problem statements that interest you at the top of sheets of flip chart paper, one statement per sheet. Attach the papers to the wall for idea gallery.

Walk around the room and write ideas directly on the papers. The ideas that accumulate on the papers will trigger other ideas as you wander around. Such movement often helps creative thinking. Ask other people to contribute ideas. Expect unexpected ideas.

Hang flip chart paper in the hall outside your office or laboratory. Write one how-to problem statement on the top. Ask each passersby to write ideas on the flip chart paper.

**• Free Word Association Imagery.**

Choose a vibrant problem statement. Intuitively select a dynamic word from that statement. Then forget the problem. Write a one-word free association to the chosen word. Then write a one-word free association to the new word. Continue writing a one-word free association to each word generated for at least six successive words.

Intuitively select one of the words. Close your eyes, and spend a few minutes forming images around the chosen word. After a few minutes, list the images on paper. Repeat this process with another word.

Then force exotic and bizarre combinations between the images and your chosen problem statement. Make them impractical, absurd, outrageous, and utterly weird.

Finally, use these bizarre image combinations to trigger practical ideas. Non-evaluatively list these ideas on flip chart paper and develop them into proposals that spark sensible, workable solutions for the original problem statement. Improve the ideas. Repeat the process. Expect unique ideas that sparkle.

---

ANOTHER FORCED-COMBINATION: Take an imaginary trip to Africa or Venus and bring back something absurd to combine with the problem at hand. Also use imaginary objects from Asia or Mars as triggers to spark unexpected ideas.

---

**• Idea Card.**

Finally, sit quietly for about 30 to 40 minutes and write one idea per card on 5" x 8" index cards with a dark marker. Stay non-evaluative as you write your ideas. Use the Zen-like principles found in non-evaluative listing and automatic writing. Occasionally write absurd, bizarre, exotic ideas and use these to trigger other ideas. Expect many unexpected ideas to emerge on these cards. Place the cards on tables, the floor, or pin them to a wall so you can see them all. Sort them. Idea board described in Chapter 14 jazzes up idea sorting.

---

A SUCCESS STORY: Idea card combined with automatic writing is superbly suited to enhance idea generation when working alone. My friend, Vicki Bradley, went quietly berserk after she learned this procedure, non-evaluatively writing every idea she had on index cards, pieces of paper, napkins, etc. A very creative artist, musician, and craftsperson, she claims idea card increased her creative output many fold. I believe her.

---

**• Combining Logical and Bizarre Ideas.**

Combine idea gallery with idea card to create a variation of combining-ideas teams described in Chapter 13. Allot at least 45-60 minutes for this procedure.

In the first 15 minutes, create bizarre and silly ideas only; non-evaluatively list these on paper. In the next 15 minutes, generate only logical solutions that make sense; non-evaluatively list these on 3" x 5" index cards, one idea to a card.

Finally, combine the ideas of the two lists and develop new, unexpected ideas. Expect it to work. It will.

**• Dream Interruption Brainwriting.**

Robert Muylaert, a manager in the Information Services Division of Federal-Mogul Corporation, shared this procedure with me.

Outline the problem using no more than five sentences. Put a pencil and a pad of paper next to your bed. At bed time, review the outline. Tell yourself that you will identify solutions to the problem, that you know you have the solution somewhere in your mind, and that you will wake yourself up when everything fits together. Muylaert claims you will probably wake up with fertile, unexpected thoughts. Jot them down at once or you will forget them. This procedure works best when you anticipate great outcomes.

As Bob Muylaert says: "This approach works well when I want something rich. I could easily use a quick fix, but by doing it this way, I get a lot of depth, some unconventional ideas, and many trigger-ideas. I have to really want to do it. It does not work if I just want to do homework. I have to enter into this procedure with anticipation and wonderment of impending greatness that will somehow be realized before morning."

**• Insert other techniques into this sequence.**

Look over the creative thinking techniques in Chapters 11-14. Stay creative as you adapt them to your use when you work alone. Brainwriting circles and clustering work especially well when you work alone. So do metaphors, random word trigger-ideas, and future fantasy.

When you finish generating ideas, convert them into a quality solution; use forced withdrawal and combine ideas into trigger-proposals (step 5 in Chapter 13), identify the criteria for a quality solution (step 6), and convert your trigger proposal into a quality solution for your organization (step 7). Working alone and achieving quality solutions adds enjoyment to work, a topic discussed in the next chapter.

Apply this sequence of creative thinking techniques to your workplace. Define a problem that continually recurs in your work, a sure sign you have not defined it effectively enough to ensure that you focus on the right problem. Use the sequence of idea-generating techniques described above. Allot enough time at work so you can do justice to creative thinking. Seize opportunities working alone.

Make action plans to do these things periodically (step 8).

---

A TRUE STORY: There's More Than One Way To Think Creatively In R&D

What is your creative thinking at work like? I am fascinated by the things creative people say about their own creative processes. My hunch is that if we were perceptive enough and had detailed insights into how each of us thinks creatively, we would discover numerous creative processes, many of which you could learn to enhance your creativity at work. Consider the following example:

I met a very creative inventor and engineer, and the then CEO of a thriving company in Michigan that has benefitted greatly from his creative inventiveness. I was so impressed by his perceptions of his creative processes, I asked him for a written description. Here's what he wrote…

"Dear Ed:

"You asked me to send you an explanation of how I think creatively. I have concluded there are two ways.

"The first creative process seems to occur almost naturally and spontaneously, without preparation, in home, work, social, or play situations. For some reason my mind set is to look for for alternatives to the accepted. In almost every situation I seem to assume that a better way exists; and I automatically find myself searching for it. My method of search involves mental exercises of turning things around, looking for an element of surprise, and allowing my subconscious to spend a few fleeting moments on it. Although I do not look for jokes at these times, they sometimes pop up.

"The second creative process is used to solve specific problems. These are design or engineering problems on products or parts of products. Preparation is required and I invest time looking at other products, reading trade periodicals, studying patents, and

---

generally observing processes and mechanisms. Such preparation loads my mind with information, most of which is not eminently useful, but pays off in the long run. I like to think of my data bank as a three dimensional grid into which I take conscious, and, I believe, subconscious forays.

"To begin my creative process I have a specific design or product need in mind, and, I almost always have a pencil and an eraser in hand. When I am alone I like to think in a darkened environment and I try to sketch the problem while making forays into my data base grid. My grid is littered with other forays from similar problems. These previous pathways, though almost all have led to dead ends, help me to penetrate my data bank, and give me a basis from which I can branch to new areas. This trial and error process, often requiring 20 to 50 attempts, is lengthy, but it is not in the least tedious.

"The potential excitement of finding a solution makes the process fun. Indeed, most of my lifetime kicks have resulted from elegant solutions extracted this way. Interestingly, the solutions may not come during an attempt. Rather, it often comes when my mind is relaxed enough that my subconscious can get my attention long enough to regurgitate the solution. I try to carve out time for quiet periods, and sometimes encourage my subconscious to react by floating an overview of the problem into my mind. Driving or sitting quietly are good times for surfacing new ideas.

"Once on the surface, the idea can generally be improved by using the first method described above, as there are a lot of weird arrangements and no humor deep in my subconscious mind."

Compelling insights. These fascinating glimpses into a creative mind at work truly impress. Notice the perception that there are two creative processes, one for daily, everyday creativity and one to solve engineering problems at work. The former involves no preparation and is very spontaneous. The second needs extensive preparation and a special focused effort. Notice also how he prepares his mind for creativity and seeks darkened environments to help.

Do you have insights into your own creative processes and triggers? Send them to me through my website: http://www.creativitybook.net

---

### • CHAPTER 20 •

### FIND AN R&D JOB YOU LIKE AND YOU'LL STAY CREATIVE

### NURTURE THE CREATIVE FLAME WITHIN YOU WHILE YOU WORK ALONE

"The enjoyment of creative thinking provides its own reward."

---

Many activities help stimulate creative juices. For example, some people respond to music, while others need absolute quiet when thinking creatively. These conditions range from mild to seemingly outrageous behaviors. When these conditions exceed the level of tolerable low-conformity at work, coworkers exert pressure on the person. Often, the person stops the behavior. In any case, creative thinking stops.

---

A TRUE STORY: In one of my creative thinking workshops, a participant told of going to his office on a Sunday morning, a rare event for him, and finding the person who shared his office working intensely with obvious relish at his desk in his underwear.

The coworker explained that he had discovered earlier in life that he worked best and stayed most creative while working in his underwear. So he came to work on Sunday when he could work unbothered by others.

---

## SELF-MOTIVATING CATALYSTS

When I tell this true story, people report this type of behavior too bizarre and unacceptable, so I offer other motivating catalysts to ponder. Most come from "Stimulating Creative thinking" by M. Stein (1975).

• Emile Zola worked at midday in artificial light.

• Lommenais worked in a room of shadowy darkness.

• Kipling wrote only with the blackest ink he could find.

• Ben Johnson performed best drinking tea while stimulated by the purring of a cat and the strong odor of orange peel.

- Schiller kept rotten apples in his desk and placed his feet in cold water.
- Shelley and Rousseau worked bareheaded in the sun.
- Boussuet worked in a cold room with his head wrapped in furs.
- Milton, Descartes, Leibnitz and Rossini lay stretched out.
- Tycho-Brache and Leibnitz worked secluded.
- Proust worked in a cork-lined room, Carlyle in a noise-proof    room.
- Balzac wore monkish working garb and worked at night with strong black coffee.
- Churchill, Frost, D'Annunzio, and Farnol worked best at night.
- Picasso painted best when someone else was in the studio.
- Guido Reni painted and de Musset wrote poetry best when dressed in magnificent style.
- Mozart worked best following exercise.
- Wagner composed music best while stroking velvet.
- A writer of Jackie Gleason's TV show had his most creative moments in the bathroom.

Some people prefer sharpened pencils, a cleared desk, a disorderly room, quiet, music, or noise. Although motivating catalysts provide needed security, you may not carry out borderline behaviors fearing what other people think.

> A TRUE STORY: I get at least one good, unexpected idea while taking a shower, and that I stay much more creative and productive if I read and write in bed as soon as I wake up without interacting with people. In fact, I wrote this book while I stayed in bed every morning and started writing when I woke up, often without stopping through mid-afternoon. I listened to Puccini's operas, and stayed most creative.

Have you discovered your own motivating catalysts? If so, can you arrange these conditions at work? If you pay attention to your motivating catalysts, you might be able to make small changes that lead to great increases in your creative output.

> A TRUE STORY: A manager of computer programmers in a Fortune-500 company told me that he gets his best ideas in the middle of the night. To make sure he captures these ideas, he has a small light, pen and paper next to his bed so he can write them down. Many people tell me similar stories. Some people tell me they get their best ideas

when driving a car, some while taking a shower or a bath, some shortly after waking in the morning, and some when walking to work. Do you know when you get your best ideas? Do you arrange for these occasions to occur more frequently to help your creative outcomes?

## FOCUS ON YOUR DAILY ENJOYMENT, NOT THE LONG-RANGE REWARDS

Research has shown that R&D people solve problems in a more creative way and turn out work with more creative surprises, if they focus their attention on their daily enjoyment and fun that comes from the challenge and their total immersion in the work (trancing out). For high levels of creative output, you need to grasp the novelty in your work, enjoy your competence and self-direction, and feel that you engage in play, rather than work. You need to have a sense that you work for your own satisfaction on a self-discovered problem in which you have considerable choices, especially in how to accomplish goals. For R&D creative thinking to flourish, you need to feel a lot of curiosity and interest, as well as have a high stability of employment to cushion taking risks.

These conditions do not usually exist, and most R&D people wait for the organization to provide the ideal workplace that never appears. Do not wait.

Immunize yourself now against the spoilers of your creative thinking. Immunize yourself against distractions, the external reward and punishment systems, evaluation and time pressures, competition with others, high control by others, and restricted choices. Keep your focus on your daily enjoyment, the challenge, and your sense of competence about your work. Nurture the creative flame within you by focusing your attention on these inner motivators. Do it now.

We all want rewards: salary raises, bonuses, promotions, awards, honors, and the like. We have to achieve organizational goals, meet deadlines, get positive performance evaluations, and obtain approval of others. Indeed, most of us constantly work for someone else's satisfaction.

Yet, these outside motivators spoil daily creative output by overwhelming inner motivation: the daily enjoyment, challenge, and self-satisfaction. Help your creative thinking by focusing your attention on inner motivators. Become as self directed as your work allows and watch your R&D creative output and innovation soar.

## ALLOW THE ENJOYMENT OF CREATIVE THINKING TO MOTIVATE YOU

Before my workshop on creative thinking techniques, I ask participants to fill out a questionnaire so I can fine-tune the workshop to their specific needs.

About 450 people in R&D, marketing, and manufacturing in six Fortune-500 companies responded. To appreciate their answer to one question, please write your response before reading further:

**"When I am creating, I feel…"**

Almost all respondents recorded good feelings. They used words like excited, fulfilled, joyful, good, enthusiastic, insightful, stimulated, enjoyable, intense, fun, happy, delighted. They wrote that staying creative made them feel satisfied, useful, energetic, alert. Other answers included challenged, worthwhile, energized. Less than 3% listed negative feelings, such as feeling anxious, frustrated, timid, stressed, disturbed, bothered, mainly because of anticipated negative reactions from colleagues.

Use these comments to help your creative output. First, you can increase enjoyment and satisfaction in your work by staying creative as a problem solver on the job.

Second, the sheer enjoyment of creative thinking provides a personal reason to stay creative. Creative thinking provides its own enjoyable rewards.

Stop allowing others in R&D to distract your attention with long-range external motivators like salary raises, bonuses, and promotion. Of course, you want these long-range rewards. Still, focus your daily work on your inner motivators, the instant enjoyment and fun inherent in creative thinking.

A HABIT THAT SPOILS R&D CREATIVE THINKING: We allow long-range **rewards** to distract us from inner motivators, our good feelings and the immediate enjoyment as we create. We allow **external** motivators to overpower and destroy **inner** motivation.

---

## • CHAPTER 21 •
## TECHNIQUES TO SELL YOUR R&D PROPOSAL CREATIVELY

Do you have trouble enrolling other R&D people in your company in your new ideas and proposals? Then expend some time and thought about how to sell it effectively. Many of the people who review your proposal have an habitual automatic No and love to give quick negative criticism. So put your proposal in a form that demonstrates you carefully evaluated and developed the proposal before presentation. Stay creative.

### EVALUATE YOUR R&D PROPOSAL FIRST

First examine your proposal's merits. Consider effectiveness, feasibility, acceptability, and difficulty.

#### • Effectiveness.

Identify the R&D problem and list your proposal's short and long-term advantages. For example, identify how your proposal will contribute to profits or improve work flow, working conditions, quality of the product or service, and methods of operation. Determine whether your proposal provides a temporary solution and whether it either partially or completely solves the problem (see 'Contingency Analysis' below).

Consider your proposal's disadvantages. Identify possible ways in which the proposal might fail and develop ways to avoid or reduce these potential problems.

#### • Feasibility.

Consider whether your proposal can coexist with current R&D policies, techniques, and objectives. Determine whether the necessary resources exist for successful implementation. Identify what new materials or processes, training programs, and changes your proposal requires. Estimate the cost of implementation, and the amount of time needed to present, gain acceptance for, and implement your proposal.

---

• **Acceptability.**

Write a clear, one page statement of the proposal that most people would understand and accept. Consider what factors might prevent people from accepting your proposal. While effectiveness and feasibility involve tangible matters, people often use intangible factors such as opinions, values, and feelings.

• **Difficulty of Implementing.**

The more difficult to implement your proposal, the greater the amount of time you must spend selling it, the larger the number of people you must persuade to accept it, and the more presentations that you must conduct in the selling process.

## SELL YOUR R&D PROPOSAL CREATIVELY WITHOUT THREAT

Request that you want people to help develop your proposal, not reject it. Resistance to change rises when someone perceives a new proposal as a threat to security and status. For example, R&D people may resist a decision to start a training program for creative thinking techniques because it threatens old ideas and the comfortable status quo. Some may react with hostility because they fear they will not easily master or use new skills. Or they may believe your proposal threatens their status or questions their performance. To overcome this, make a list of everyone who might feel threatened by your proposal and how to reduce the perceived and real threat. Include how your proposal benefits each person.

Some R&D people may even reject a new proposal because they did not develop it. Therefore, determine in advance:

• where your proposal can motivate rather than threaten.

• who would object.

• how to emphasize benefits and needs.

• the impact on people's personal or financial status.

• where difficulties to understand or implement exist.

• where it affects existing professional R&D relationships.

• whether it generates new challenges or responsibilities.

• how to develop such opportunities.

Review your proposal's advantages and disadvantages. Prepare answers to questions that may arise regarding the disadvantages. Make changes that assist in reducing the disadvantages.

## PRESENT YOUR PROPOSAL CAREFULLY

Start your presentation with a general description, and then follow with a detailed explanation. For example, when you make a proposal in R&D to conduct workshops teaching creative thinking techniques, the general description might include a description of creative thinking techniques and the reason to do it: to enhance profits by developing new products, patents, procedures, processes, or services. The detailed explanation might specify the content of the workshops; their dates and times, whether the workshops will occur immediately throughout the company or gradually by work groups and departments; and who has the responsibility for implementation.

If your proposal involves changes in techniques, authority, or responsibilities, consider how such changes will affect the people present. This, in turn, determines the overall approach of the presentation. People reject promising proposals because of an inappropriate approach when selling them.

Present your proposal in an understated manner. Avoid a hard sell approach. Over-enthusiasm, especially at the beginning of the presentation, can have a negative effect. In addition, avoid jargon or technical terms unless people understand them.

To correct defects in your presentation, rehearse before friends or colleagues. Even a minor flaw or omitted point can result in quick negative criticism followed by rejection.

Visual aids clarify technical aspects and help people retain material. Recall rises with visual aids. Reinforce main points and benefits to key people. Keep your visual aids simple. Each chart or graph should cover only one idea. Unnecessary or poor visuals create a more negative impact than no visual aids at all.

## RESOLVE CONFLICT CREATIVELY IN R&D

Conflict becomes inevitable when selling a proposal, since everyone has a different viewpoint. Used effectively, conflict improves proposals by stimulating new ideas. However, it must be handled carefully or it can impair successful implementation.

Therefore avoid statements that lead to unnecessary conflict, such as stereotyping or pigeonholing people. On the other hand, openly discuss potential adversarial conditions based on different value systems or different needs.

There are four basic approaches to conflict resolution: win-lose, lose-win, lose-lose, and win-win.

**Win-lose**:

In win-lose conflicts, you achieve your goals at the expense of others. Authority rule occurs when you promote your own interests without soliciting support from others. Majority rule occurs through voting when you expect a favorable vote. Minority rule occurs when you impose your ideas on others with the aid of a small, but influential group.

Use win-lose when you feel strongly about a particular proposal, when a one-time transaction does not involve a long-term relationship, or during a crisis when there is no time for consulting others.

**Lose-win**:

With this method, you give in to suggestions of others. Use this method to obtain needed support and when you want harmony more than winning.

**Lose-lose**:

This method involves compromise in which no one gets what they want; for example, using arbitration in which a third party arrives at a solution possibly detrimental to both sides, and during tradeoffs in which one side agrees to a number of points in exchange for concessions on other issues. Use lose-lose when no agreement seems possible otherwise.

**Win-win**:

Use this method to manage conflict when you want all sides to feel they have won and treat the proposal as their own. Resolve conflict through consensus negotiation that achieves commitment to the proposal from all parties. See T. Gordon (1977) "Leader Effectiveness Training (L.E.T.)" for a classic account of this process.

In win-win, focus the obstacles to implementation on problems, rather than personalities. Solve the problems creatively with mutual agreements. Win-win succeeds when more time is first spent identifying the obstacles and defining problems than immediately seeking solutions (see Chapter 11).

Use win-win to (a) gather maximum input from everyone involved; (b) gain wide support during implementation of the proposal; and (c) develop long-term cooperation. By enlisting participation early, you can persuade others to accept the proposal as their own. The win-win method consumes time well spent, since it results in a high degree of satisfaction and commitment to the proposal. This usually generates support from all involved.

You can help your proposal succeed by soliciting deficiencies and possible disadvantages from others, and then requesting help so faults can be corrected before implementation. When you enlist other people to plan and formulate the proposal, fewer problems and criticisms arise later.

The win-win method requires that you listen carefully, since comments from others will contain useful information. Respond by paraphrasing major points and summarizing what others have said.

An open attitude helps. When you push through a proposal, or appear annoyed with suggested alternatives, you can lose the support of others. Instead, incorporate the suggestions of others in your proposal whenever possible to insure win-win. Although win-win usually provides the best approach, successfully selling proposals requires learning to identify the appropriate method for the situation and using that method in a skilled manner.

## CONTINGENCY ANALYSIS

Analyze your proposal from the viewpoint of positives and negatives; what you like about it and what can go wrong; weaknesses. deficiencies, improvements needed; and blocks and barriers during implementation that other people would use as reasons to reject your idea.

Then non-evaluatively list how to overcome or reduce the important negatives.

| POSITIVES | NEGATIVES |
|---|---|
| List what you like about your proposal. Also list its strengths. | List everything that can go wrong. Also list weaknesses, deficiencies, improvements needed, blocks & barriers.    (Put all in the form of "How to" problem statements.) |
|  | How to... |
|  | How to... |
|  | How to... |
|  | How to... |
|  | How to... |
|  | How to... |
|  | How to... |
|  | How to... |
|  | How to... |

Non-Evaluatively List Ways to Reduce the Impact of Negatives on Your Proposal:

# • PART 6 •

## TECHNIQUES FOR LEADERS OF R&D TEAMS

"You can't afford to let your competition get more creative than you, can you?"

--------------------------------------------------------

### • CHAPTER 22 •

## YOU DO WANT YOUR R&D PEOPLE TO BECOME MORE CREATIVE, DON'T YOU?

"To be around butterflies, you have to generously help caterpillars."

*"The biggest help to my creativity is when my boss leaves town ..."*

A TRUE STORY: I am engaged in a research project related to enhancing applied creativity at work. Among the many items in a questionnaire, I ask people to list "the biggest help to their creativity at work." The results are unexpected. The biggest response by far, about half, is "other people." Items like time, challenge, and freedom occur at a much lower frequency. Rewards are hardly mentioned. Conspicuously absent are customers and vendors. One person did respond "when my boss leaves town."

Much food for creative thought here.

As an R&D leader, help the people in your work group think more creatively. Perhaps your work group wallows in too much complacency or has a routine way of doing business. If the following conditions exist, you may want to boost creative thinking:

• R&D team members resist trying out something new.

• Everyone thinks alike....or thinks very little.

• People do the same things in the same ways.

• People lack interest in what others are doing.

• Performance levels stay constant.

• Your R&D team lacks risk-taking.

• The atmosphere does not challenge enough.

• People ask few challenging questions.

Many R&D leaders resist new creative thinking techniques. My discussions with people in Fortune-500 companies lead me to the following reasons for many of their doubts. Which fit you?

• Some R&D leaders see creative thinking as a mysterious ability and talent of a gifted few. This belief produces a negative attitude toward learning or using these techniques.

• Some R&D leaders believe we inherit creative thinking skills. Thus, they think that a highly creative person needs no help, and that a low creative person stays beyond help. Easy to see how this thinking leads to resistance.

• Some R&D leaders think that learning and applying creativity techniques a waste of time (and money). Since they believe this, they don't use them to keep the work group on the creative track.

• Some R&D leaders think their work group already creative enough. "Too many good ideas lying around," they say. They think their work group needs more doers, more implementers, more product champions, more entrepreneurs, etc. Thus, they don't use creativity techniques.

• Some R&D leaders really do want a more innovative work group, but doubt that creative thinking techniques help, usually because they do not have enough experience or information about them. This book can help them change their mind.

• Some R&D leaders believe that only one right way to do anything exists and they cannot understand all that fuss about thinking creatively. Their pessimism and negative attitude becomes a self-fulfilling prophecy. Hence, no creative thinking techniques for the work group.

• What keeps you from using creativity techniques to boost creative thinking with your team?

---

• CHAPTER 23 •

## TECHNIQUES TO MOTIVATE R&D PEOPLE

### NURTURE THE CREATIVE FLAME IN OTHERS

"There's a great deal of fun in being creative.
Not so much HA-HA fun as AH-HA fun."

---

Feelings control motivation, the desire to do something. R&D people often carry out activities that seem to offer nothing beyond the activity itself. No surprises here. You do this with your hobbies or sports. How much people engage in an activity at work without a tangible reward reflects their inner motivation; the real reward lies in the good feelings (fun, enjoyment, pleasure, delight, joy) produced by doing the activity.

Inner motivation stimulates creative accomplishment at work if people find it satisfying and enjoyable. Sometimes people become so focused they don't notice their surroundings. They trance out and report feeling elated. Elation provides an inner reward in hobbies, games, and aesthetic experiences, but it doesn't occur often enough at work.

Some of the fuel that produces daily enjoyment in R&D includes: feeling competent and self-directed; working independently on a challenging problem; sense of satisfaction; getting daily respect; total immersion in work; a sense of play rather than work; learning new things. These provide a boost to creative work in R&D.

On the other hand, motivators from outside a person produce negative feelings: tenseness, fear, resentment, irritation, or worse, due to the usual reward and punishment systems that R&D leaders use. Negative feelings interfere with and lower the quality of R&D creative work, the reason we consider their effects here.

These outside negative motivators for creativity at work include expected rewards (salary raises, promotions, bonuses, recognition, awards, honors) and punishments (not getting rewards, penalties for failure, performance appraisals, evaluations, imposed deadlines, competition with others, high control from others, restricted choices, close supervision and intrusions, watched by others, red tape, assigned routine jobs, meeting imposed goals, lack of freedom, abundance of quick negative criticism, little

---

encouragement or appreciation, low acceptance of ideas, ineffective meetings, inappropriate leadership styles, unsuitable rewards, limited resources, overload of work, interruptions, demands of others, emphasis on productivity rather than creative thinking, limited communication, mountains of paperwork).

As an R&D leader, to increase the desire to produce creative outputs, make work inner-rewarding. Unfortunately, a huge snag exists; external motivators easily overwhelm and replace inner motivators. In other words, when R&D leaders externally reward or punish people for creative activities that usually depend on self-motivation, the motivator shifts to the outside reward or punishment, and inner motivation declines. While R&D productivity on routine approved projects may rise, self-motivated creative outputs fall.

Paradoxically, external rewards that provide positive feedback on competence leads to high self-motivation, but not if people perceive the feedback as manipulative or controlling. Continual exposure to controlling rewards and negative information about competence lowers the desire to produce creative work because it spoils people's self-motivation.

The same applies to **praise**. If a person perceives praise as feedback on competence, then praise increases inner motivation. However, if the person perceives praise as a reward, and then works to get the reward, resentment increases and creative output declines because self-motivation drops.

So remember: external rewards have this dual effect. They can provide positive feedback about a person's competence and increase self-motivation for creative outcomes. Or they can control behavior and lower self-motivation, thereby spoiling creative work.

Still, self-motivation and creative output can increase if you focus on the daily enjoyment in the work itself, instead of on getting the reward.

Probably you, as R&D leader, don't think of many rewards except salary, bonuses, and promotion at work. Still, inner rewards exist in almost any situation: provide clear feedback; make sure the work has a broad range of choices and challenges; delegate effectively and expect self-direction. If you do this, any activity becomes enjoyable and internally rewarding.

Even though external rewards lower performance on interesting tasks, such as solving problems creatively, the same external rewards improve performance on routine, mundane tasks. Indeed, you may need external rewards, even coercion, to get people to do certain jobs. But the more you build self-satisfaction into boring activities, the more people stay self-motivated, creative, and productive at work.

## SOME RESEARCH ON THE ENJOYMENT OF CREATIVE THINKING

Before my workshop on creative thinking techniques, I ask participants to fill out a questionnaire so I can fine-tune the workshop to their specific needs.

About 450 people in R&D, marketing, and manufacturing in six Fortune-500 companies responded. To appreciate their answer to one question, please write your response before reading further:

"When I am creating, I feel…"

Almost all respondents said they had good feelings: they used words like excited, fulfilled, joyful, good, enthusiastic, insightful, stimulated, enjoyable, intense, fun, happy, delighted; they wrote staying creative made them feel good, satisfied, useful, energetic, alert; other answers included challenged, worthwhile, energized. Less than 3% listed negative feelings, such as feeling anxious, frustrated, timid, stressed, disturbed, bothered, mainly because of anticipated negative reactions from colleagues.

Learn from these comments to lead your R&D team more effectively. First, encourage enjoyment and satisfaction in work to help people stay creative as problem solvers on the job and boost creative outcomes.

Second, use the sheer enjoyment of creative thinking to provide an inner reward for your R&D people to stay creative. Help people focus their daily work on the instant enjoyment and fun inherent in creative thinking.

One manager told me about an R&D leader who greeted his people with: "Are you having fun today?" If they answered yes, he asked them to share the fun with him. If they answered no, he asked what he could do to help them have fun.

Use the approach: "Are you having fun today?" to focus people on their inner motivators and the desire to stay creative: the good feelings, the enjoyment and the fun that people report feeling when they create.

So if you want to stimulate creative work, stop distracting the attention of your R&D work unit with long-range external motivators, like salary raises and bonuses and promotion. Of course, you cannot do away with these rewards. But you can help people focus the daily work on the instant enjoyment inherent in creative thinking.

---

A HABIT THAT SPOILS R&D CREATIVE THINKING: Leaders allow **long-range rewards** to distract people from inner motivators, the immediate enjoyment of producing creative outcomes.

---

## TECHNIQUES TO MOTIVATE FOR CREATIVE OUTCOMES:
## MORE RESEARCH

Self-motivation fuels creative work, while outside motivators sink it. Factors that help productivity usually spoil creative effort, including offers of rewards, expected evaluation, imposed deadlines, and other conditions that common wisdom proclaims helps and benefits performance. Some research supports this notion (see "The Social Psychology Of Creative Thinking" by T. Amabile and "Creativity In The R&D Laboratory" by T. Amabile & S. Gryskiewicz):

• A high interest in an activity improves creative outcomes.

• Focusing on inner reasons (enjoyment) to do something helps creative thinking, while focusing interest on outside reasons spoils creative thinking.

• Doing an activity for its own sake (fun, enjoyment, elation, pleasure) helps creative thinking, while doing something to accomplish an outside goal spoils creative thinking.

• Creative thinking spoils when people focus on expected evaluation, a reward that depends on performance, supervision, lack of choice to do something or how to do it, or imposed deadlines.

• Creative thinking increases when self-motivation rises due to the enjoyment that comes with choices, no obvious intrusions, no expectation of evaluation, and no supervision by watchers.

The outside motivators shown to spoil creative outcomes include:

• Focusing on external rewards rather than on inner rewards: the enjoyment and fun at work.

• Thinking about outside reasons (rewards and punishments) for doing an activity.

• Evaluation, or even expecting evaluation.

• Reduced choice of what to do and how to carry it out.

• Watched while performing an activity.

• Expecting an attractive reward for good performance.

• Externally imposed deadlines and time constraints.

As an R&D leader, help your work group self-motivate for creative work. Try this:

• Match the activity to worker interest and involvement.

• Encourage self-direction.

• Stress inner reasons for doing things (daily enjoyment, joy, fun, excitement, pleasure).

• Help people reduce the feelings (resentment, fear) generated by visible external constraints.

• Offer some free choice about whether or how to do an assignment.

• Avoid obvious intrusions and minimize performance evaluations.

• Make work self-rewarding so it increases the desire to produce creative outcomes.

---

A HABIT THAT SPOILS R&D CREATIVE THINKING: Leaders focus on **external** rewards to motivate people at work and devalue the **inner** reasons to accomplish something, the daily enjoyment and excitement at work.

---

## ATTENTION R&D LEADERS

Much research shows that people turn out work more creative and solve problems more creatively if they focus their attention on their daily enjoyment and the challenges of the work.

As an R&D leader, encourage your work group to achieve high levels of creative output: help them perceive the novelty in the work, their own self-competence and self-direction, and the sense they engage in play rather than work. Help them feel that they work for their own satisfaction on a self-discovered problem in which they have many choices, especially in how to do the work. For creative work to flourish, help them feel a lot of curiosity and interest.

If you want the people in your R&D team to increase creative output, do not distract them from these sources of self-motivation. Do not dangle external motivators in front of them on a daily basis. Instead focus their daily attention, and yours too, on inner motivators: the daily enjoyment due to the novelty and challenges of the work, their sense of competence, self-direction, and satisfaction.

Give them as many choices as you can, especially in how to achieve goals. Provide job stability to encourage risk taking, the core of creative enterprise at work. Allow them to feel they engage in play rather than work. Encourage self-evaluation, self-direction, and self-satisfaction at work.

Allow R&D people to focus on their inner motivators, not on outside motivators. They all want salary raises, promotions, and honors, all important rewards. The R&D people in your work group must accomplish your goals, meet deadlines, get positive performance appraisals, and obtain the good will of others. They do work for your satisfaction. Yet, these outside motivators spoil creative enterprise by overwhelming inner motivators: the daily enjoyment, pleasure, and satisfaction of the work, the elation that comes from achieving self-determined goals.

Help creative thinking by focusing R&D people's attention on these inner motivators. Help people in your work group become self directed by adjusting your leadership style and watch creative work take off in your work group.

## IMMUNIZE AGAINST THE SPOILERS

Research has shown that R&D people solve problems in a more creative way and turn out work with more creative surprises, if they focus their attention on their daily enjoyment and fun that comes from the challenge, and their total immersion in the work (trancing out). For high levels of creative output, people need to feel they engage in play, rather than work. They need to have a sense that they work for their own satisfaction on a self-discovered problem in which they have considerable choices, especially in how to accomplish goals. For creative thinking to flourish, they need to have high stability of employment to help shift paradigms and take risks.

You may find it difficult to provide these conditions at work. Most people wait for the organization to provide the ideal workplace that never appears. Do not allow your R&D team to wait.

Help them immunize themselves against the spoilers of creative thinking: the distractions, the external reward and punishment systems, evaluation and time pressures, competition with others, high control by others, and restricted choices. Keep their focus on their daily enjoyment, the challenge, and their sense of competence about the work. Nurture the creative flame within them by focusing attention on their inner motivators now.

They all need rewards: salary raises, bonuses, promotions, other awards and honors. They have to achieve your goals, meet deadlines, get positive performance evaluations, and obtain approval of others. Indeed, they do work for your satisfaction.

Yet, these outside motivators spoil daily creative output by overwhelming inner motivation: the daily enjoyment of creative effort. Help their creative thinking by focusing attention on inner motivators. Immunize now. Allow them to become as self directed as the work allows, and watch their creative output soar.

# EXTERNAL AND INTERNAL MOTIVATORS IN R&D
To help you understand external and internal motivators, look at these examples.

| EXTERNAL MOTIVATORS | INTERNAL MOTIVATORS |
|---|---|
| 1. Working on something to meet an external goal. | 1. Working on something for its own sake. |
| 2. Concern with expected external evaluation. | 2. Self-evaluation. |
| 3. Desire for recognition from others. | 3. Self-recognition prevails. |
| 4. Focusing on outside competition. | 4. Self-competitive to do better than before. |
| 5. Reaction against external time pressures and deadlines. | 5. No time pressures or deadlines. |
| 6. Focusing on visible constraints. | 6. Absence of external constraints. |
| 7. Material gain, promotion, status, affiliation with others, self-enhancement, self-defense, etc. | 7. The daily enjoyment in the process, the joy of solution, and the total immersion in the work. |
| 8. Perception that they work to obtain some external goal. | 8. Perception that the work comes from one's own interest. |
| 9. Wanting to attain a reward, meet a deadline, achieve approval, obtain positive evaluation. | 9. The work has novelty; provides a sense of competence and self-direction; and has deep involvement and playfulness. |
| 10. High financial and conceptual control from superiors. | 10. Low financial and conceptual control from superiors. |
| 11. None of these =>. | 11. The individual stays curious; stimulated by the activity; gains a sense of competence; perceives the activity as free from external control; and has a sense of play rather than work. |
| 12. Being watched. | 12. Not watched. |
| 13. People perceive evaluations and rewards as externally controlling and manipulative. | 13. People perceive evaluations and rewards providing information on competence. |
| 14. Honors, if seen as controlling and manipulative. | 14. Occasional honors, if perceived as indicating competence. |
| 15. Meeting demands of others; lack of self-direction. | 15. Self directed work. |

| **EXTERNAL MOTIVATORS** | **INTERNAL MOTIVATORS** |
|---|---|
| 16. Working on an uninteresting activity. | 16. Working on an interesting activity. |
| 17. No choice of whether to do the job or how to do it. | 17. Free choice, especially in how to carry out the job. |
| 18. Problems chosen by others. | 18. Self-discovered problems. |
| 19. No curiosity. | 19. Curiosity about the activity. |
| 20. Working for someone else's satisfaction. | 20. Working for self-satisfaction. |
| 21. Depending on superiors. | 21. Working independently. |
| 22. External criteria for determining success or failure. | 22. Internal criteria are used. |
| 23. Focusing on external rewards. | 23. Focusing on internal rewards. |
| 24. Others initiate activities. | 24. High responsibility for initiating new activities. |
| 25. No power to hire or choose assistants. | 25. High degree of power to hire or choose assistants. |
| 26. Much interference from superiors. | 26. No interference from superiors. |
| 27. Job instability. | 27. High stability of employment. |
| 28. Low immunity to the perception of external motivators. | 28. High immunity to external motivators. |

# IN SUMMARY

### • EXTERNAL REWARDS ...

• decrease the activity without the reward

• decrease enjoyment and undermine performance

• shift attention away from inner motivators and the activity

• imply work rather than play

• lower risk-taking and the complexity of a chosen activity

• decrease interest in previously interesting activities

• spoil creative enterprise

### • INNER REWARDS ...

• increase the likelihood people will perform the activity later

• enhance performance

• direct attention toward the activity itself

• imply play rather than work

• increase realistic risk-taking and complexity of a chosen task

• increase interest

• help creative enterprise

---

**• CHAPTER 24 •**

**"THE BIGGEST HELP TO MY CREATIVITY AT WORK IS ...**
**WHEN MY BOSS LEAVES TOWN."**

---

A TRUE STORY: I am engaged in a research project related to enhancing applied creativity at work. Among the many items in a questionnaire, I ask people to list "the biggest help to their creativity at work." The results are unexpected. The biggest response by far, about half, is "other people." Items like time, challenge, and freedom occur at a much lower frequency. Rewards are hardly mentioned. Conspicuously absent are customers and vendors. One person did respond "when my boss leaves town." Much food for creative thought here.

---

A TRUE STORY: In Chapter 23, I described a research project. Before attending my workshop on creative thinking, I ask the participants to fill out a questionnaire so I can fine-tune the workshop to specific needs. The responses of about 450 people in R&D, marketing, and manufacturing in six Fortune-500 companies reveal a great deal. To appreciate their comments, please jot down your own response before reading further:

"When I am creating, I feel…"



These respondents used words like: excited, fulfilled, joyful, good, enthusiastic, insightful, stimulated, enjoyable, intense, fun, happy, delighted. They wrote staying creative made them feel good, satisfied, useful, energetic, alert. Other answers included challenged, worthwhile, energized. Less than 3% listed negative feelings, such as feeling anxious, frustrated, timid, stressed, disturbed, bothered, mainly because of anticipated negative reactions from colleagues.

I find these comments important for two reasons. First, you can increase enjoyment and satisfaction in your work by staying more creative as a problem solver on the job.

Second, the sheer enjoyment of creative thinking provides an important, personal reason to stay creative. Creative thinking provides its own enjoyable rewards. Stop allowing others to distract your attention with long-range external motivators like salary raises, bonuses, and promotion. Of course, you want these long-range rewards. Still, focus your daily work on intrinsic motivation, the instant enjoyment inherent in on-the-job creative thinking, and watch your creative output soar.

Let us continue with responses to other statements in my pre-workshop questionnaire. Please jot down your own responses to:

**"The biggest help to my creative thinking at work..."**

**"The biggest obstacle to my creative thinking at work..."**

**"I need the following from my job environment to stay more creative..."**

The main responses by 93 R&D chemists and engineers to the statement in my questionnaire: "The biggest help to my creative thinking at work is..." including the actual numbers of R&D chemists and engineers who mentioned this item:

| "THE BIGGEST HELP TO MY CREATIVE THINKING AT WORK IS...." | |
|---|---|
| | R&D Chemists & Engineers |
| a. Support and encouragement from other people | 25 |
| b. Sharing ideas with other people | 17 |
| c. Time | 14 |
| d. Challenging task; adventurous feelings | 11 |
| e. Freedom | 8 |
| f. Being alone | 8 |
| g. Perceived a problem to be solved | 7 |
| h. Rewards, credit, acknowledgment | 5 |
| i. Miscellaneous | 9 |

Forty-two (combined items a and b above) of these R&D chemists and engineers (almost half) mentioned interaction with other people as the biggest help to their creative thinking at work.

Surprisingly, freedom did not rank first (item e), and eight people listed 'being alone.' (item f). One conclusion: R&D leaders need to treat people differently so all types become self-motivated to stay creative. Adopt the attitude that everyone gets special treatment at work.

I found similar results in a Fortune-500 company involving 20 R&D managers, 32 R&D supervisors, and 24 of their subordinates who responded to "The biggest help to my creative thinking is....." in this way (see the Table below):

| **"THE BIGGEST HELP TO MY CREATIVE THINKING AT WORK IS...."** | 20 R&D Managers | 32 R&D Supervisors | 24 Scientists & Engineers |
|---|---|---|---|
| Support and encouragement from other people | 4 | 5 | 5 |
| Sharing ideas with others | 6 | 7 | 12 |
| Supervision | 0 | 0 | 4 |
| Time | 6 | 11 | 2 |
| Challenge | 1 | 1 | 0 |
| Freedom | 5 | 9 | 4 |
| Being alone | 1 | 1 | 1 |
| Perceived problem to be solved | 0 | 0 | 0 |
| Rewards, recognition, acknowledgment | 0 | 0 | 0 |
| Overcoming personal limitations | 1 | 3 | 0 |
| Resources | 0 | 0 | 2 |
| My internal resources | 0 | 0 | 4 |

(The totals exceed the number of people, since some people wrote more than one thing.)

Again no one listed rewards. They wrote support, encouragement, and sharing ideas with other people as the biggest help; they also listed time and freedom. The similarity between these and the responses above strikes me as significant.

Lest you think only R&D personnel respond this way, I present a summary of how 54 non-R&D people responded: Other people (36); Time (7); Climate (8); Challenge (3); Freedom (4); Miscellaneous (4). This sample includes mostly managers and professionals in marketing, human resources, and manufacturing from 34, mostly Fortune-500 companies. Again no one mentioned rewards.

I categorized responses from 24 managers in three large International Fortune-500 Companies in England as: Other people (12); Freedom (6); Challenge (2); Time (2), Climate (2); Understand job (2); Rewards (1); Miscellaneous (3). Though few in number, these people appear similar to managers and professionals in the United States.

Interestingly, no one mentioned vendors or customers as the biggest help to creative thinking. Not using customers in this way does not fit the stated goals of these market oriented firms. Use vendors to help creative output since vendors do want to help.

One R&D scientist actually wrote: "The biggest help to my creativity is ... when my boss leaves town." Was he writing about you?

Overall, these results indicate that many R&D people perceive that other people provide the biggest help to their creative thinking. This provides an important clue on how team leaders can spur creative output.

First: Encourage and build in activities so R&D people interact more frequently, exchanging and discussing each other's ideas.

Second: Bring in an occasional expert professional to help people's ideas.

Third: Arrange for more creative thinking in small groups during regular R&D meetings. For example, make it a norm that "buzz groups" define and generate ideas to help solve at least one problem in every meeting of your work group, as described in Chapter 12 of this book. This will help the exchange of ideas and enhance creative thinking.

**"The Biggest Obstacle in My Job Environment To My Creative Thinking..."**

What blocks creative thinking at work? A few people mentioned personal limitations, newness to the job, lack of creative thinking skills, etc. However, almost all the comments from the 450 respondents included conditions you can control: lack of time;

lack of freedom; abundance of quick negative criticism; distractions; low encouragement; low acceptance of new ideas; ineffective meetings. Some wrote about cautious management styles; red tape; lack of appreciation; unsuitable rewards. Others blamed limited resources; overload of work; interruptions; demands of others; the need to stay productive rather than creative; limited communication; mountains of paperwork.

The good news: you can correct most of these in the workplace.

### "I Need The Following From My Job Environment To Be More Creative..."

What stimulates on-the-job creative thinking? The 450 respondents concentrated on: more talking to others; more time; more freedom; less red tape, paperwork and routine jobs; better resources; more respect as a professional; more recognition for innovation; better communications; an atmosphere that encourages originality; fewer meetings; better teamwork; fewer penalties for failure; fewer interruptions; more supportive atmosphere.

As one R&D person wrote: "the opportunity to be heard, openness, more participation in selection of assignments, more freedom in selection of approaches, less daily and weekly accounting of activity."

If you want to find out what spoils creative thinking or how to stimulate creative output in your R&D team, ask your people either directly or through questionnaires. Create your own questionnaires. If this process overwhelms you, obtain the help of consultants who can facilitate the process and ensure favorable outcomes.

### ENCOURAGE INDIVIDUALIZED MOTIVATING CATALYSTS

Review the motivating catalysts that help stimulate creative juices in Chapter 23. Catalytic triggers take different forms. Some people prefer sharpened pencils, or typewriters, or a cleared desk, or a disorderly room, or quiet, or music, or noise. Some people will not carry out borderline behaviors fearing what other people may think. When the motivating catalyst exceeds the level of tolerable low conformity in your R&D work group, isolation of the person may occur and creative output declines.

Are you spoiling potential creative output in your R&D work group by not tolerating low conformity? Pay attention to motivating catalysts and you might make small changes in the workplace that greatly helps creative thinking in your work group.

## WHY HIGHLY CREATIVE R&D PEOPLE DON'T ALWAYS PRODUCE HIGHLY CREATIVE OUTPUTS

Ideally, leaders want R&D people with creative abilities to produce creative outputs. Unfortunately, this gets sidetracked when risky new ideas threaten the security of colleagues. This leads to overt or subtle negative feedback to the idea-person, whose sense of job security becomes threatened. Self-motivation declines. The result: less creative thinking; less new and risky ideas; worsening relations between the R&D idea-person and colleagues; or the idea-person quits or retires on the job, becoming a weekend creative.

R&D people with high creative ability do not always produce highly creative outputs. The outcomes also depend on the specific job conditions and on an atmosphere that helps creative output. Help creative R&D people resist and immunize themselves against the spoilers of creative thinking. Still, the workplace and its climate need to be in line to spur creative outcomes. Otherwise, self-motivation for creative work declines and dies.

In addition, not being able to communicate new ideas to others spoils self-motivation for creative R&D effort. This happens mainly in **older** projects with well-developed ideas, rather than in new projects seeking new creative ideas. Thus, the creative person helps a newly formed work group, but at later stages, hurts productivity, and, in turn may become hurt when the mature team overwhelms inner motivation. Because a highly coordinated, goal oriented R&D team strives to achieve its mission, it spoils creative thinking with its lack of flexibility.

A non-flexible R&D workplace produces disappointment and resentment in a creative person, especially a low conformer. Self-motivation for creative thinking drops. This results in the creative person cutting down on creative outcomes or leaving the job. Often most self-motivation disappears and the creative person retires on the job, becoming what I call a 'weekend creative,' channeling creative energies into weekend pursuits. This outcome doesn't benefit the R&D team or the company. Recapture these weekend energies for creative outputs at work.

TRUE STORY: I have met many R&D weekend creatives, full-time employees of Fortune-500 companies, who focus their creative energy on weekend pursuits. These include an artist, a sculptor, a manager of six restaurants, an owner of an oriental antique shop, and a designer of houses. These people engaged in these engrossing activities in addition to their full-time job. If any people in your R&D team act as weekend creatives, don't you want to capture their energies for creative thinking in your work group? The most famous weekend creative: Albert Einstein, who perfected his theory of relativity while working as a clerk in the patent office in Bern, Switzerland. Many others exist.

## TECHNIQUES TO BOOST SUBMISSION OF IDEAS IN YOUR WORK GROUP

A creative R&D person may hold back ideas in a non-supportive workplace because the ideas push against mainstream thinking or imply criticism of others. Or, other people attack the ideas with quick negative criticism. Or the idea may not get to resource providers because people pigeonhole the person and do not take him or her seriously.

Also, R&D people have a personal conflict about how to spend time to obtain desired rewards. Clear rewards exist for good results on approved projects, but the rewards for new ideas become vague. People don't submit ideas if the rewards expected from idea submission do not justify the time lost on an approved project.

So R&D people may not invest enough time to fully develop an idea because of pressure from current projects. Incomplete ideas prevent effective evaluation by management, who rejects the idea. This creates the "I don't believe it" attitude when leaders announce policies to reward new ideas and innovations.

To get out of this difficulty, emphasize new ideas and innovation in R&D job descriptions, objectives, performance reviews. Include self-motivation.

Clarify what you want in new ideas. Provide help from other people to submit a fully developed idea. Reward the idea originator with the opportunity to implement the idea, if desired.

To increase creative output from R&D idea people, provide the idea person high independence, job security, and training in selling ideas in a non-threatening way. An R&D idea person needs room to act on his or her ideas and hunches, rather than tightly organized work. An R&D idea person needs new projects with flexibility; as older projects become more productive rather than innovative, goals become inflexible, and creative effort interferes.

The creative R&D person will do poorly if you don't appreciate his or her ideas, because resentment increases, self-motivation dies, and creative enterprise goes down. The creative R&D person needs influence or a champion to get ideas accepted and resources provided. Otherwise, self-motivation dies, stifling creative work.

Thus, creative output falls when a potentially creative R&D person lacks confidence, status, training, and security; lacks communication with critical resource providers; and lacks flexibility in goals and approaches. Self-motivation for creative output drops.

Enhance the creative payoff for potentially creative R&D people. Keep the work coordination low, but not too loose. Management may tell **what** to do, but not how to do it. Enable the creative R&D person to directly influence important decision makers to get approval for resources and encouragement. Provide ongoing opportunities for discussing new ideas with others, since many R&D people say other people provide the biggest help to their creative thinking at work. Allow people to self-motivate for greater creative output. Provide a leaky system with many types of opportunities to obtain approval and resources to implement ideas and projects.

## REDUCE THE SPOILERS OF YOUR COMPANY'S INNOVATION PROCESS

Some R&D work groups do not require much creative thinking. They want experts who solve problems by well-established techniques. If so, management rewards productivity, but not originality. The R&D person with a really new approach may not have a chance to report it, obtain resources, get higher status, or respect. People stop thinking creatively and innovation slows if the workplace does provide encouragement.

As a R&D leader, if you need innovative ideas to solve a problem, help creative thinking by encouraging people to use advanced creative thinking techniques. Ask R&D people what they need. Determine the flexibility of their work situation. Encourage new projects, new areas, or an additional challenging project. Arrange it so coordination of work does not deter adventurous R&D people. Increase the opportunities for them to try out ideas on others and target appropriate rewards for their efforts. Many rewards go to R&D high conformers who stay productive along well-established paths, while R&D low conformers miss out on rewards because people often misunderstand their ideas.

What if you do not want much creative thinking? Or if you want to turn it on and off? Encourage creative idea generation. Then set firm limits afterwards on the type of ideas you want when identifying the **criteria** to select ideas. Do not set the limits or the criteria to select ideas **before** idea generation. This spoils creative thinking.

In other words, encourage R&D people to get their ideas and imagination up into outer space, and then gently help engineer them back to earth. This produces a large mixture of ideas from which to choose. Excellence results from this approach.

---

### • CHAPTER 25 •

## SELF-DIRECTED TEAM BUILDING IN R&D WITHOUT A CONSULTANT

"You do want your team to become more successful, don't you?"

---

TRUE STORY: A Fortune-500 company asked me to present a workshop combining creative thinking techniques with team building for 30 R&D engineers. I presented a modified 2-day workshop on creative thinking techniques, in which we worked on work related problems, followed by a 1-day workshop on "self directed team building without a consultant," customized especially for this event. The results: startlingly effective, especially since everyone thought these people resistant to team building.

## YOU HAVE NO CHOICE;
## YOU MUST BUILD YOUR R&D TEAM

Do you want to spend time building your R&D team for excellence? You really have no choice, because you build your team all the time. The issue: you either build an effective, successful team, or you build something else. Whatever you do, your R&D team responds. Take care. Everything that happens in your R&D team, and all the qualities that you like and dislike in your team, results from your actions. You have 100% responsibility. Why? Because you can lead your team to perform better and achieve more using self directed team building principles, or you don't. You can't avoid this responsibility.

How do you lead your R&D team? As a 'star with helpers' or as 'first among equals'? Many R&D team leaders like to think they act 'first among equals' but often discover they pursue a 'star with helpers' approach. Remember: the best approach remains a flexible one that depends on the situation. Still, if you want a more "first among equals" approach, use this self-directed team building sequence that **doesn't** require a consultant.

FIRST: Set aside time at the end of each meeting to discuss the quality of the team's interactions and the quality of the results. Don't accept the glib way people say they liked it. Focus on what went well and what needs improving. Who dominated? Who said

---

nothing? Who helped creative effort? What hindered creative thinking? How can the team make meetings more creative?

If you sense you create some of the problem, ask others for ways you can become part of the solution. It often helps if the chair rotates among your team members, a key step for you to become "first among equals."

SECOND: Ask everyone in the meeting to rate in writing on a scale of 1 to 5 (with 5 the highest) the following issues:

|  | LOW | HIGH |
|---|---|---|
| • Participation in the discussion equally balanced among all members? | 1 2 3 4 5 | |
| • Your opinions and thoughts solicited by the team? | 1 2 3 4 5 | |
| • You influenced the outcome? | 1 2 3 4 5 | |
| • Others influenced the outcome? | 1 2 3 4 5 | |
| • Creative outcomes? | 1 2 3 4 5 | |

Ask each person to write a one sentence comment about each response to the questions above. In addition, ask everyone to write down behaviors they found helpful to the team effort and behaviors that hindered the team. Collect the ratings and comments at the end of the meeting and provide a written summary to everyone. Discuss this summary openly at the next meeting of your team.

THIRD: Whatever your approach to leading your R&D team, do it with excellence. Welcome opportunities to be "first among equals" and a "star with helpers." Develop skill to use either style depending on what you need to get the job done.

## WHO BUILDS THE R&D TEAM?

Does the team leader, the team, or the consultant build the R&D team? Many people think the consultant builds the team. Not so! YOU, the team leader, builds your team. You avoid this responsibility at your peril. Don't copy the R&D team leader who left town the day I guided his team through a team building session. "Go ahead without me," he said. He became a major part of the problem.

Self directed R&D team building means the team leader learns how to build his or her team without a consultant, an important role of all team leaders. Usually team leaders hire consultants to guide them in building their own team. Unfortunately, the signals get crossed, the team leader takes a back seat, and the consultant builds the team. Don't allow this to happen. The R&D team leader must build the team.

If the leader needs a consultant, use one to help the team in a conventional way. This includes the following steps: getting the agreement of all R&D team members; interviewing each member; reporting feedback to the team; and helping the team develop action plans committed to excellence, increasing creative outputs, achieving better team performance and success. A complicated process, it takes a lot of time and costs a lot of money. But first emphasize self-directed team building and do all you can to build your R&D team without a consultant.

## WHY PEOPLE DON'T SUPPORT R&D TEAM DECISIONS:
## SILENCE DOESN'T MEAN CONSENT

How your R&D team makes a decision affects creative outputs. Often an R&D team moves in an important direction thinking everyone agrees, only to find out later that some people have serious reservations that results in a lack of teamwork and cooperation. Why did apparent consensus turn into subtle sabotage? Part of the misunderstanding comes from the effects of decision making.

Consensus does not mean an unanimous vote. Consensus indicates a decision that the majority favors, and the minority, after having ample time to persuade and influence the majority, agrees to implement and publicly support as the best decision possible for this R&D team as a whole. Consensus often takes a great deal of time to achieve.

Consensus insures that the truth that lies in the minority surfaces and gets a hearing. In addition, differences of opinion produce additional data and new perspectives, further clarify the issues, and forces the R&D team to seek better information and make a broader-based higher-quality decision.

R&D teams often make decisions in other ways because they find consensus hard to achieve and time consuming, thereby insuring lower teamwork and low cooperation.

• **PSEUDO-CONSENSUS**: a decision where the R&D team encourages and interprets silence as consent, or where members pressure each other to orally agree when they do not. This results in implied support which disappears at crucial times.

• **MAJORITY VOTE**: a common form of decision making in R&D teams, assumes the minority will go along willingly with the majority. Often it does. Still, people sometimes resent the action and give only token support during implementation.

Voting often creates coalitions that focus on how to win the next vote, not on how to best implement the decision. Voting also fosters clever arguments, rather than a rational, balanced discussion that leads to a cohesive team cooperating to implement the best decision for that team.

• **MINORITY WINS**: a decision where a powerful minority ramrods a decision which the majority does not support. As expected, this leads to little future support.

• **HANDSHAKE**: a decision where a suggestion produces instant permission to continue from another person. This puts the team into action before determining whether the team backs the decision. Eventually, support lags during implementation.

Cliques within a R&D team use handshake, a short-lived method to control an entire team. It cannot exist when all team members voice their opinion, consider everyone's opinion, and keep the team aware of how it makes decisions.

• **SELF-AUTHORIZATION**: a decision where a R&D team member makes a suggestion and immediately continues assuming that since no one objected, everyone agreed. Again, silence implied consent. Even if the team agrees, members may resent the way the decision emerged.

• Two hurtful ways your R&D team may make decisions: The **FIZZLE**, where team members totally ignore a suggestion, and the **CHOP**, where one or more powerful members of the R&D team immediately reject a suggestion. Both these actions produce a climate where people carefully filter ideas, suggestions, and opinions. This results in the silence interpreted as consent.

• **TRUE CONSENSUS**, the antidote to these toxic decision-making modes, does not always lead to the most elegant, brilliant, exciting, or dashing decision. But consensus does frequently produce the most supported, most implementable plan that leads to cooperation, cohesiveness, and teamwork among the members of your R&D team.

## R&D TEAM INTERACTION PATTERNS; WHO TALKS MOST AND TO WHOM

A TRUE STORY: I stumbled onto the usefulness of team interaction patterns when I served on a committee I found particularly frustrating. I disliked those meetings, and I couldn't figure out why.

I asked one of my colleagues what he thought. All he said was, "I wish he (the team leader) would stop talking so we could talk." Yet we both agreed that the team leader did not talk that much. Rather his timing threw us off. He followed an interruption pattern in which he always made a short comment whenever anyone else said anything. This stopped free discussion.

So I asked the team leader how he viewed the meetings. "It's like pulling teeth to get that bunch to talk," he said.

How remarkable. We wanted to talk, and he wanted us to talk, and yet we all allowed an unproductive team interaction pattern to stop an open discussion.

I asked him what he would prefer, and he described an ideal interaction where each team member talks freely in order to move the team closer to its goals.

Asked how we could achieve this pattern, he said that he would stop talking. He realized that he was partly responsible for what happened. I suppose in his mind an interactive would emerge if he stopped talking.

When he did stop talking, a long silence ensued. Finally someone asked if he felt okay. At his insistence, the discussion continued without his usual short comments. Very quickly the old interruptive pattern appeared, only this time another person took over the role of team leader. Apparently we felt comfortable with this dominant team leader pattern.

Thus I learned the power and effects of Team Interaction Patterns. Actually, these patterns produce good results in the right situation. Problems occur when an R&D team uses one pattern almost exclusively or uses a pattern at the wrong time.

Teams use many patterns:

• Spray And Pray: one person talks in a long-winded monologue, and hopes, in the absence of feedback, that people remain interested and understand the monologue.

• Tinkers to Evans to Chance: three people form a dominating clique and exclude other people from the discussion.

• Endless Dialogue: the team leader and one person engage in extended dialogue, often with little apparent substance.

• The Outsider: one team member uses silence, doodling, reading, and facing away from the team center to show lack of attention.

The interaction patterns of your R&D team will tell you a lot. Do your team's interaction patterns prove useful? Do they move the team toward desirable goals or do they waste valuable time of your team? Worse yet, do you contribute to the problem?

Share this chapter with your R&D team and discuss it at the end of a meeting. Ask everyone to identify at least two team interaction patterns that occurred during the meeting and assess their helpfulness. And discuss ways to improve the interaction patterns of your team to help achieve your goals and improve team effectiveness. Consider this an important part of self directed team building and one way to help your R&D team move toward creativeness and excellence.

## SET R&D TEAM NORMS THROUGH OPEN DISCUSSION

Most R&D teams unawarely set norms for behavior by accepting unproductive and interfering behaviors of its members. Occasionally, team members challenge a particular behavior, but usually R&D people tolerate in silence even though silence does not mean consent. A team's acceptance of unacceptable behavior can have a profound negative effect on the team's performance.

To counteract this particular poison, set team norms through discussion of issues that team members bring up. Ask your R&D team to write ways the team could respond to the following situations:

1. A team member does more than his or her share of the work.

2. A team member does less than his or her share of the work.

3. A team member talks too much.

4. A team member talks too little.

5. A team member comes to meetings unprepared.

6. A team member comes late or not at all.

7. A team member does not listen or interrupts other people.

8. A team member ignores team goals and wanders off the subject.

9. A team member horses around and clowns to excess.

10. A team member or clique dominates the meeting.

11. A team member, hostile and aggressive, puts other people down and kills their ideas.

After each member of the R&D team writes his or her thoughts, ask each person to share ideas on the issues that impact them the most. Do not ignore any issue however trivial it may appear to you.

If the situations listed above do not fit your team, ask them to volunteer situations that do apply to them. Lead your R&D team to real consensus on what to do about these situations in the future. Don't blame your R&D people; no one told them about their counterproductive actions until now.

These discussions produce important outcomes. First, team members learn productive norms of the team. Second, the R&D team develops ways to handle unproductive situations in the future.

Prepare for disagreements. Negotiate these creatively in a win-win way. See T. Gordon (1977) Leader Effectiveness Training (L.E.T.) for creative ways to resolve conflict. Or hire a consultant to help.

## FEEDBACK WITHIN THE R&D TEAM

One extremely helpful norm relates to open feedback for R&D team improvement. To help develop this norm, ask team members to discuss the following questions immediately after a team meeting:

1. Balanced discussion: Did the team hear and use the ideas of everyone? What can the team do to improve?

2. Examine contributions and ideas sufficiently: Did the team clarify, amend, adopt, or ignore peoples input? Who dominated? Who helped?

3. Freely expressed agreements and disagreements: Did the team use them in the discussion? How did the team help creative thinking?

4. Was the controversy person-centered or problem-centered?

5. The creative atmosphere: How can the team improve it?

6. Contribution to the final team outcome: What helped you to contribute? What hindered?

7. Our creative thinking: How can we to improve it?

8. Feelings about the meeting: What did you like about this meeting? Did people stay creative enough?

Open discussion on the norms used by your R&D team contributes to your team's ability to work together, increase performance, stimulates creative effort, and improves results, the goal of self directed team building.

## SHARE THE GOALS OF SELF DIRECTED R&D TEAM BUILDING

Make R&D team members aware of the goals of self directed team building, so they can help. Consider the following goals:

• A non-threatening team atmosphere: Since R&D members often expose ignorance, take risks with new ideas, and engage in controversy, truly productive discussions depend on an open, accepting atmosphere.

• Cooperative effort and a teamwork: R&D members feel free to ask for help without stigma, and give it when needed. Counterproductive behaviors, such as ridicule, one-upmanship, sarcasm, or status seeking do not exist. Controversy results in creative win-win solutions to problems.

• Accomplishing team goals accepted as the R&D team's primary purpose: Team members recognize high level accomplishment as the main point of the existence of the team. If this does not happen, productivity lags and excellence does not happen.

• Every R&D member participates in discussions: Silent members may get a lot out of the discussion, but they do not contribute to the team effort.

• Leader and follower roles distributed among the R&D team's members: A few members should not usurp responsibility for productivity, creative thinking, and quality. Everyone learns and performs leader and follower functions.

• Enjoyable and pleasant R&D work sessions: Controversy becomes a spark to trigger enjoyment and involvement in a discussion.

• Feedback and evaluation, an accepted norm: R&D team members review and discuss problems openly to improve team excellence.

• Behaviors fundamental to good relationships within an R&D team: Team members attend regularly, come prepared, and keep agreements with each other.

• R&D Team excellence: Quality outcomes lead to success and depend on interdependence, a common language, challenge, trust and respect, creative thinking and problem-solving skills, confrontation and conflict-resolving skills, and a desire to celebrate the large and small successes of the team.

---

## • CHAPTER 26 •

## CREATIVITY, CONFORMITY, AND RISK

## TECHNIQUES TO MANAGE LOW CONFORMERS IN R&D

---

R&D creative thinking, generating new ideas on the forefront, involves risk: risk of failure; risk that people will laugh, make fun, humiliate, and even denounce as ridiculous. So the extent that R&D people will conform to mainstream thinking determines how much they willingly risk. These traits, high or low risk taking, high or low conforming, determines the kinds of creative outcomes your R&D work group creates.

R&D work groups insist on some level of conformity to norms, rules, customs, and policies. **High** conformers define problems and generate ideas along conventional paths, preferring solutions that conform to existing conditions. They make their creative output compatible with work norms and do not shift paradigms easily. Usually uncomfortable with the bizarre, they do not rock the boat. They solve problems narrowly. High conformers do things better, not differently.

**Low** conformers, the opposite of this, prefer to shift paradigms broadly, and explode problem definitions beyond conventional norms and reasonable boundaries of the original problem. They love the bizarre and enjoy rocking the boat, sinking it if they can. They solve problems innovatively. Low conformers do things differently, not just better.

---

A TRUE STORY: I once observed a R&D work group tackling a problem of absenteeism that exceeded acceptable levels in the plant by just a few percentage points. Some of the members, clearly **high** conformers, suggested sending letters to frequent absentees, providing rewards for non-absenteeism and penalties for excessive absenteeism, all within the bounds of absenteeism.

On the other hand, other members, clearly **low** conformers, suggested very different things. For example, they shifted the paradigm as the real problem became motivation, not absenteeism. We have to motivate people to come to work. How? Paint and decorate the building. Everyone gets an office with a window. Build a gym. They suggested a new

---

building. In a new town. How about in Paris, France? How about more fun at work? More parties. Less work, more fun? No work, all fun. If top management will not go along, fire them (much laughter). All this triggered by absenteeism that was a little higher than desired.

Does this seem familiar? Two opposing types in conflict over style: each style affects how people shift paradigms; define problems; and generate, select, and develop ideas; in the acceptance of change itself.

A TRUE STORY: A vice-president of R&D, a confessed low conformer, told me about a very profitable machine he developed after denied permission to build it. He found space in the back of a warehouse, built a partition around it to mimic the back wall, and hid the entrance to the hidden room with packing cases and boxes. He developed the machine in secret with a few trusted colleagues. Only after he finished it did he inform the delighted company president, the same person who had denied him permission to develop it.

Extreme high or low conformers don't exist at work, even if you think you know one. Locate yourself or other people on the scale between the extremes? The following may help:

R&D **high** conformers, reliable and efficient, tend toward precision. They solve problems in conventional ways and don't shift paradigms very much. They seek stability. Other people see them as safe and dependable. They put in long hours on detailed work without boredom and challenge rules only with strong support from others. They comply. High conformers, sensitive to people, work to maintain team cohesion, teamwork, and cooperation. On a scale of 1 to 5, do these traits fit you or others?

Do high conformers prove useful in a work group? Of course. Without them, chaos occurs, and little gets finished. An organization without enough high conformers falls apart, since high conformers provide the necessary glue that holds things together.

R&D **low** conformers, on the other hand, tend to lack discipline. They shift paradigms frequently and approach problems from new angles. They challenge a problem's assumptions and define a problem incessantly. They don't comply easily. Their indifference to team consensus leads people to see them as abrasive, undependable, impractical, or combative. Routine tasks bore them. They challenge rules frequently, and have low respect for past customs. Insensitive to other people and team cohesion, they don't cooperate. On a scale of 1 to 5, do these preferences fit you or others?

Do low conformers prove valuable? Of course. An organization without enough low conformers staggers toward complacency and stagnation. It heads for difficulty the next time the environment changes to produce a crisis.

Low or high conformance has little to do with creative ability. Equally creative within their styles, people still perceive low conformers as more creative because they take higher risks, have a more adventurous spirit, and use bizarre trigger-ideas during problem solving.

A third R&D type exists, people in the middle range of conformance, **middle** conformers, who span the gap between high and low conformers. These people can, within limits, act as moderate low or high conformers, and can communicate with both extremes. The moderate conformers of a R&D work group usually includes its leader, since leaders must be able to communicate with all people. Again, this has no bearing on creative ability, but reflects a preference for moderation.

The extreme types have much to say about each other. R&D high conformers say low conformers act weird, too difficult to work with, too undependable, and they don't want to work with them. "He took an important assignment and developed a brilliant solution to a problem I never gave him." (Interestingly, low conformers say the same thing about other low conformers, effectively isolating themselves from most people.)

In contrast, R&D low conformers say high conformers act like sticks in the mud, red tape bureaucrats, uptight, narrow-minded people who spoil everyone's creative thinking. As expected, low conformers say they do not want to work with high conformers.

Is it any wonder that work groups require middle conformers to manage them? As an R&D leader, help low and high conformers respect each other's unique contributions. Get them to recognize each other's value and understand that each type does necessary jobs that the other dislikes. The high conformer loves digging in one place, while the low conformer abhors it. The low conformer loves to leap around, nibbling here and there, looking for the best place to dig. Not so the high conformer. Thus you need both types to succeed. And you need to act like a moderate conformer to manage them.

> A TRUE STORY: After I discussed this theory about low and high conformers in a workshop for a Fortune-500 company, one R&D supervisor told me about a person working on a problem for over two years with no progress, probably a low conformer working on a problem that needed a high conformer.
>
> Six weeks later, I presented another workshop at the same location, and the manager of the work group told me that I had saved them a "bundle" of money. Apparently they "switched the nibbler (a low conformer) with a digger (a high conformer) and made more progress in six weeks than in the previous 2 1/2 years."

Match the job to the style of the R&D person. High conformers do best on conforming jobs, and low conformers do best on innovative jobs. Taking these personal preference styles into account increases the chances for quality solutions at work.

### R&D LOW AND HIGH CONFORMERS CAN HELP EACH OTHER

Discuss this openly. Make R&D high and low conformers aware of their importance to produce quality outputs. Make them proud of what they uniquely accomplish and contribute. After all, the high conformer provides the solid foundation for the low conformer's risky activities, while the low conformer provides the impetus for periodic change to avoid complacency and stagnation. Thus, when collaborating, R&D high conformers supply stability and continuity, while R&D low conformers supply the break with past traditions and accepted norms. Excellence results from a collaboration between both types.

### LEAD THE R&D LOW CONFORMER IN YOUR WORK UNIT CREATIVELY

Most R&D managers manage the high conformer relatively easily. High conformers work well in teams and cooperate with policies and work group norms. Low conformers require different approaches.

Since you need low conformers to escape routine complacency in your R&D work group, focus on the negotiating and delegating leadership styles with a careful sortie occasionally into the collaborative style (see Chapter 27 for "leadership styles"). With some exceptions, low conformers tend to stay loners and need special conditions to operate effectively.

The R&D low conformer, as a work group member, clashes when working with logical, linear thinkers. Low conformers suffocate in rigid cultures and cause distress to other R&D people: they perceive others as destructive to their creative efforts. They become impatient with routine repetitive jobs, and this often results in premature task termination. They irritate or alienate other people with their intense drive, their focus on pet projects, and their idiosyncrasies. Easily isolated, R&D colleagues often perceive them as self-serving loners and disruptive to efforts of the work group. Low conformers can suffer from loneliness, isolation, and feeling different.

If you, the R&D leader, act as a low conformer, you often find leadership activities uninteresting and irritating. People may perceive you as isolated, impractical, and lacking in follow-through. You may develop interpersonal barriers with people in your own work group and avoid giving important coaching and feedback and thus avoid important aspects of your job. Often unwilling to delegate, and impatient with day-to-day operating details, the R&D low-conforming leader may become impulsive and perceived by other people as erratic, condescending, and distracted. In addition, the low-conformer leader can feel superior to others and become highly dictatorial.

Every R&D low conformer exists as a majority of one. Easily bored, they would rather move into untried areas not worrying about risk or troubled by ambiguity. Uninterested in social matters, they may lack social skills. They want to use their minds to solve difficult, personally fulfilling problems. They experience their work as a calling. When working in unexplored areas, they do well without support or approval from others.

**Leading** low conformers in your R&D work group requires a special capacity for patience and good will. Some approaches make it a little easier, but do not expect miracles. Bright low conformers can overwhelm, so develop a unique style for each low conformer in your R&D work group.

• Encourage low conformers in your work group to use the motivating catalysts that stimulate their work.

• Let your low conformers help you and the work group get out of routine complacency.

• Make sure low conformers know the objectives of the R&D team so they work toward similar goals.

• Tolerate their honesty.

• Let the low conformers know that you consider them respected, valuable members of the R&D team.

• Accept the low conformer's firm stance without calling them stubborn.

• Help the low conformers in your R&D work group see the progress they have made when depressed and discouraged.

• Do not interpret the low conformer's continual dissatisfaction as disloyalty.

• Enjoy the low conformer's bizarre ideas.

• Tolerate the flow of ideas without asking the low conformers in your R&D work group to settle on one too early.

• Resist urging them to seek early solutions to problems, the quick fix.

• Do not burden the low conformers in your R&D work group with suggestions that slow them down when hot on the trail of a solution.

• Keep the R&D facilities low conformers need open at any time.

• Accept the low conformer's independence and do not take offense when they go ahead without asking your advice.

• Provide rewards that you can tailor to their choice.

• Help prevent them from wearing other R&D people out.

• Interact with low conformers without formality.

• Get their input, but don't try to make them a team player.

• Accept the low conformer's fantasies and do not accuse them of being unrealistic.

• Help low conformers sell their proposals and make them relevant.

• Respect their periods of isolation.

• Help low conformers learn from their failures.

• Give them encouraging feedback concerning their ideas.

• Help provide the kinds of people interactions they need.

• Help set up R&D work structures that accommodate their style.

• Help low conformers feel secure so they risk creative effort.

• Support low conformers when others criticize them.

• Provide for the stimulation low conformers need, especially from other professionals.

• Give low conformers enough time for the incubation process to work.

• Provide low conformers in your R&D work group the support needed during the depressing episodes of the creative process.

• Have a two-paths-up promotion system for low conformers.

A TRUE STORY: A division of a Fortune-500 Company decided to hold a Creative Fair involving several hundred people in R&D, marketing, and manufacturing. They planned exhibits around current technologies and products, idea-generating sessions led by volunteers, and eventual presentation of ideas for new directions to senior managers. They called me in to present a one-day workshop for the volunteers, so they would spur creative thinking and quality solutions in their sessions, not spoil it. My workshop turned into a one-day problem-solving creative thinking event.

Because the volunteers tended toward low-conformer types, we explored the impact of interactions between low and high conformers, and what to do about it. We practiced four techniques they could use. We finished up generating ideas on how they could effectively ensure success using idea card. I later heard great things about the contribution of the volunteer leaders who attended my workshop.

---

**• CHAPTER 27 •**

**ADJUST YOUR LEADERSHIP STYLE TO HELP**

**CREATIVE WORK IN YOUR R&D TEAM**

---

### TO THE LEADER OF THE R&D TEAM

Your R&D work unit can provide you with useful data on how you impact their creative work. Small adjustments in your leadership style can result in a large improvement in their effectiveness, greater motivation for them to stay creative, and an increase in the creative output and innovations of your work group.

Your leadership style includes the behaviors you use to influence the people in your R&D work group to accomplish goals.

Your real style probably eludes you; the people in your work group can help you recognize it. To determine your real style, *find out how they perceive your leadership interactions and how you affect their creative efforts*. Ask them to respond in writing. After you read their input, respond to your work group in a way that...

(a) increases their trust in you

(b) improves the overall creative outcomes of your work group for excellence and creative output

(c) raises their motivation to stay more creative

(d) improves your leadership flexibility and effectiveness.

For example, you might call a meeting of your R&D work group, thank them for their cooperation, reveal what you learned, and then plan ways with them to improve your leadership interactions so you improve the quality of the creative climate, raise the creative output of the work group, and improve the quality of the interactions between all of you.

### YOUR LEADERSHIP STYLE

Most R&D leader interactions fall into two types: **task** behaviors, which involve directive, one-way communications explaining what, when, where and how to do a task, and **supportive** behaviors, which involve two-way communications, non-evaluative listening, stroking, and encouraging behavior. The relative frequency with which you combine these two types of interactions constitutes your leadership style. You can combine these interactions into four useful leadership styles, each with an array of effective behaviors:

Style I. **DIRECTIVE** STYLE: Your predominant behaviors: telling, asserting, and modeling.

Style II. **PARTICIPATIVE** STYLE: Your predominant behaviors: coaching, negotiating, and collaborating.

Style III. **CATALYTIC** STYLE: Your predominant behaviors: encouraging, facilitating, and consulting.

Style IV. **NON-DIRECTIVE** STYLE: Your predominant behavior: delegating.

Choose your leadership interactions according to the ability, willingness, and confidence of the R&D person to work independently. That is, he or she has the willingness to do the task, has the ability, has a high performance level, and has the confidence he or she can accomplish the task. Use the directive style if the person has low abilities to operate independently. Otherwise, use the leadership styles that exert less control.

## ADJUST YOUR LEADERSHIP STYLE

You will find it difficult to change a leadership style. Yet, to help your R&D team become more creative and productive, you must become more flexible and use the skills from all four styles. How can you do this? Add enough skills one at a time so you can respond flexibly to the people in your work group.

For example: To become a more flexible R&D leader who encourages creative effort, you might learn how to **assert** for a more effective directive style; how to **coach** and how to **negotiate** disagreements for a more effective participative style; how to **listen** and respond non-evaluatively for a more effective catalytic style; how to **delegate** for a more effective non-directive style; and, of course, how to manage and **motivate** for creative work.

You can adjust your leadership style by combining task and supportive behaviors to boost the creative output of your work group.

R&D leaders who want to find out what spoils innovation and creative outputs for excellence, or how to stimulate creative thought, need only to ask their work group, either directly or anonymously through questionnaires, or with the help of consultants who can facilitate the process and ensure favorable outcomes.

Some people will respond they want more time to stay creative and freedom of choice: time to think creatively; freedom to choose what to work on; freedom to decide how to accomplish goals; and the time to do it. Some caution, however.

First: Farris[1] reported that the most innovative N.A.S.A. teams, as judged by senior management, included mainly those whose supervisors' styles remained **moderate**, neither too tight nor too loose.

Second: Andrews and Pelz[2] found that scientists in industry who were judged most effective by others in terms of scientific contributions and usefulness to the organization, had **moderate** controls by others, neither too tight nor too loose.

---

[1]Farris, G. F. (1973) The Technical Supervisor. Technical Review 75 (5), April Issue.

[2]Pelz, D. C. and F. M. Andrews (1976) Scientists In Organizations. Institute for Social Research. University of Michigan.

---

In other words, when we focus on the best leadership style for highly innovative outcomes judged by external standards of appraisal, complete freedom of choice does less well than **moderate** freedom combined with supportive consultations with supervisors or managers. The catalytic and non-directive style, mixed with a very careful use of the participative style, would seem the most useful to stimulate creative output and innovation of your work group.

Thus, research indicates that over the long run, the leadership habit of exercising too little control works less well than the **middle** ground.

A HABIT THAT SPOILS R&D CREATIVE THINKING: R&D leaders do not **adjust** their leadership style to meet the needs of their work group for creative work and innovation.

**R&D leaders have a host of habits that interfere with creative effort.**

These include:

- not expecting people to stay creative;
- rationing resources rather than seeking new creative opportunities;
- not tolerating ambiguity or loose ends, and moving to closure too quickly;
- over-emphasizing reason and logic, coupled with a belief that they waste time with fantasy and intuition;
- using inappropriate strategies to solve problems;
- not allowing creative people to individualize their working conditions;
- emphasizing external rewards and goals as motivators, rather than the inner reward of daily enjoyment;
- failing to allot time for ideas to incubate in the mind and mature in the hidden complexities of creative thinking;
- not questioning how to improve things;
- playing it safe and not taking risks;
- not allowing bizarre ideas to exist in the early stages of problem solving;
- dominating meetings;
- not providing training in using advanced creative thinking technologies;
- not making it clear that they want creative thinking;
- committing to an early idea too soon;
- using quick negative criticism to stifle an idea or a paradigm shift;
- insisting that a new, fragile idea run the gauntlet of the work group's negative thinking before they take it seriously;
- not allowing the time needed to shift paradigms and solve problems creatively.

Many R&D leaders remain **unaware** that their own habits, otherwise productive, contribute so directly to their work group's low creative output.

To increase the creative outcomes of work groups, R&D leaders must form the habit of cycling leadership styles: setting direction, giving a clear idea of the end product they want, delegating within the person's areas of expertise using the non-directive leadership style, and then taking the risk of letting people set their own goals and run their own business, with occasional encouragement and support using the catalytic style.

## HELP R&D CREATIVE OUTPUT THROUGH RESPONSIBLE SELF-DIRECTION

Self-directed activity with high inner motivation helps creative efforts. **Internal** rewards for solving a problem creatively include excitement, enjoyment, interest, novelty, a sense of control over one's work, curiosity, positive feedback to oneself on competence, etc. In contrast, **external** rewards spoil creative thinking by distracting people from the daily enjoyment of creative work. These distractions spoil creative thinking. See Chapter 23 and Amabile (1983) for a good discussion of this.

Thus, one crucial issue in managing for creative outputs involves helping the people in your R&D work group move toward responsible self direction: "responsible" because they respect organizational goals; and "self direction" because they become self motivated to:

• Perform their job competently and effectively;

• Take responsibility;

• Set high and realistic goals;

• Negotiate and keep agreements;

• Solve problems creatively;

• Accept direction from others, when necessary;

• Plan and use their time wisely;

• Work productively alone and with others to accomplish goals.

## CYCLE YOUR R&D LEADERSHIP INTERACTIONS

To help R&D people become self directed and more creative, cycle your leadership interactions. Your behavior changes with their increasing ability to take charge and work independently on each specific task.

This leadership approach values four leadership styles. Some other theories value only the participative leadership style. Yet people remain less likely to become self directed and self motivated if you only use the participative style. The independence allowed in the non-directive style and the encouragement you provide using the catalytic style boosts the creative process.

## ENCOURAGE PARTICIPATIVE INTERACTIONS IN R&D MEETINGS

When you dominate R&D meetings, you hinder people from solving problems creatively and from becoming more self directed. Participative interaction in meetings will occur when you and your work group use effective group discussion skills, collaborate instead of compete, use consensus decision making, rotate the chair, and periodically review and discuss ways to improve while solving problems during meetings. If your R&D work group lacks these skills, you may want to consider team excellence training for your work group. See Chapter 25 for self directed R&D team building approaches.

Research has shown that work groups trained in team excellence often produce creative solutions of high quality. These approaches focus discussion on relevant topics and help achieve consensus, while increasing team cohesiveness. Commitment to implement decisions also increases.

In contrast, untrained work groups frequently produce outcomes of lower quality. Such work groups often contain dominating individuals or cliques who pursue personal agendas rather than group goals. Use self directed team building to develop creative thinking, teamwork, and cooperation in teams (see Chapter 25).

## NEGOTIATE DISAGREEMENTS CREATIVELY

Negotiating disagreements using win-win problem-solving techniques can help the people in your R&D work group become more self directed and more creative when solving problems (see Gordon, 1977). Success depends on your ability to assert, listen for understanding, respond non-evaluatively, and mutually generate and agree on solutions with the people in your work group. If you do not possess these skills, you may want to learn them or call in a third party to help manage excessive conflict.

## USE JOB ENRICHMENT

Enrich the jobs of the people in your R&D work group through challenging tasks to help them become self directed and more creative. Do not confuse this with making changes in routine work tasks that lead to job enlargement or job rotation. True job enrichment occurs when you remove some controls; increase accountability for their own work; assign a complete task; grant additional authority; introduce more difficult tasks; or assign unique roles. Changes like these encourage more self-direction and creative thinking in your work group.

## USE PARTICIPATIVE GOAL SETTING AND PERFORMANCE APPRAISAL

If you don't share performance appraisal, if you don't provide frequent feedback, if you don't encourage R&D people to respond, then resentment may develop that curtails self directed activity and lowers creative thinking. In addition, you have lost an excellent opportunity for performance development. Gordon (1977) describes an excellent performance appraisal process involving mutual description of job roles, mutual setting of work goals, and mutual evaluation of goal accomplishment.

The skills needed by you and the people in your R&D work group to do this effectively involve asserting, active listening and responding, and negotiating using win-win problem-solving techniques. This process encourages people in your R&D work group to become self directed and more creative.

Consider reversing the usual goal setting procedure. Instead of telling your goals, ask each person in your R&D work group to state his or her goals. If you like them, back them to the hilt, and encourage creative effort. If you do not accept them, negotiate mutually acceptable goals. Creative work and productivity flourishes with this approach.

## PROVIDE TRAINING

It makes sense to provide training in the skills needed to become self-directed and to solve innovation problems creatively. People in your R&D work group must have the following skills in order to carry out organization goals responsibly with a minimum of supervision: negotiate: keep agreements; take responsibility, and set high and realistic goals and objectives; plan and use their time wisely; reduce and handle the stress in their job; work productively alone and with others in teams to accomplish goals; use advanced creative thinking techniques to solve problems. The basic skills needed to do this include teamwork and team interaction skills; time management and stress management skills; communication skills for resolving conflict creatively; and, of course, advanced creative thinking techniques to shift paradigms to achieve quality solutions and high levels of creative output and innovation..

# R&D LEADER'S GOOD RELATIONSHIPS IMPROVE CREATIVE OUTPUT & INNOVATION

**Relationships** in R&D teams matter. In my own research, summarized in Chapter 24 of this book, fully 50% of R&D scientists and engineers wrote that "other people" were the biggest help to their creativity at work. These R&D people have good relationships.

On the other hand, one R&D scientist wrote that the biggest help to his creativity is "when my boss leaves town." Clearly a poor relationship.

Along the same lines, Basu & Green (R&D Innovator, Volume 3, Number 11; 1994) found that creative output and innovation tended to increase when an R&D team leader had good relationships with subordinates.

The process of innovation often involves compromise, downplaying egos, and providing rewards fairly in the R&D team. When team leaders and members had good relationships, a climate of cooperation helped R&D success.

In other words, the creative output and innovations of R&D people tended to rise as relationships with the team leader improved. Basu & Green found that good relationships also had non-material benefits that boost researcher creativity. This includes more freedom: freedom to explore new ideas, freedom to work on personal projects, and freedom to exchange information with people outside the company to help innovation.

Successful R&D scientists & engineers reported managerial support in the form of emotional and administrative assistance for risky projects. When faced with technical obstacles, these people saw their leaders as more motivating and encouraging, quicker to act on paperwork and financial requests, and less likely to penalize failure. A strong correlation exists between good relationships and managerial support for innovation.

R&D people who have good relationships with the team leader seemed more committed to the organization. They reported more inner motivation, higher satisfaction, better attitudes towards innovation, and more willingness to achieve organization goals. And the higher levels of motivation and involvement yielded increased creative output and innovation.

This research suggests that **team building** activities that improve relationships take place on a regular basis in R&D. Chapter 25 in this book describes "Techniques For **Self Directed Team Building** To Help Creative Thinking Without A Consultant" that can help in this process. Check it out.

Basu & Green believe that when an R&D team leader has good relationships with his or her subordinates, it becomes emotionally and politically easier for scientists and engineers to produce creative output and innovations. R&D personnel in poor relationships with the team leader tended to take fewer risks, engage less in unconventional ideas, and can muster only the minimum of the resources needed for the creative process. In fact, leaders had difficulty motivating R&D people who did not perceive their leader as an ally.

Basu & Green also found that charismatic leaders create more meaningful relationships with team members, and thus stimulate the best in subordinates. This doesn't lower the importance of technical expertise, which R&D people insist is important in team leaders; its just not as essential as the ability to inspire, motivate, and energize.

Also, charismatic leaders excite and inspire. Interpersonal attraction is important in R&D and these managers created a sense of urgency among the scientists and engineers who report to them. In turn, these R&D scientists and engineers adopted more positive attitudes towards innovation. Charismatic leaders also fostered creative output & innovation by generating higher levels of commitment among subordinates towards the company.

**WHAT TYPE OF LEADER ARE YOU?**

**WHAT RELATIONSHIPS DO YOU FOSTER?**

## LEAD TO FOSTER CREATIVE WORK AND INNOVATION

Circle the number that best describes your leadership style.

|              | UNLIKE |   |   |   |   |   | LIKE |
|--------------|--------|---|---|---|---|---|------|
|              | ME     | 1 | 2 | 3 | 4 | 5 | ME   |

### I. Directive Leader:

I enjoy taking charge and getting things done. I prefer specificity and objectivity. I believe if I clarify and define the facts and techniques, then personalities and feelings should have no significant influence. I like to decide the way to do a task, and then tell the people in my work group how to do it. I have a sincere wish for my people to succeed at what they do. I generally solve problems and disagreements alone. People sometimes resent my controlling behavior. Sometimes they perceive me as dictating, dominating or overprotective.

|              | UNLIKE |   |   |   |   |   | LIKE |
|--------------|--------|---|---|---|---|---|------|
|              | ME     | 1 | 2 | 3 | 4 | 5 | ME   |

### II. Participative Leader:

I believe that hard work and learning helps the people in my work group to realize their potential. I put my energy and focus my commitment on their development as well as on accomplishing the task. I solicit their ideas and show a great interest in their work. I give frequent informal comments and coaching on their performance. I negotiate disagreements through mutual problem solving techniques. Sometimes they perceive me as compromising too soon.

|              | UNLIKE |   |   |   |   |   | LIKE |
|--------------|--------|---|---|---|---|---|------|
|              | ME     | 1 | 2 | 3 | 4 | 5 | ME   |

### III. Catalytic Leader:

I ensure that people in my work group grow in confidence and ability to perform their tasks. I recognize achievements and let them make decisions and solve their own problems. I listen non-evaluatively and encourage them with a warm personal approach. I solve problems and disagreements in a catalytic way, that is, I do not actually involve myself except as a catalyst. They sometimes perceive me as patronizing, condescending and meddling.

|              | UNLIKE |   |   |   |   |   | LIKE |
|--------------|--------|---|---|---|---|---|------|
|              | ME     | 1 | 2 | 3 | 4 | 5 | ME   |

### IV. Non-Directive Leader:

I like to concentrate on the big picture and enjoy planning the future. I delegate. I assign tasks or make requests, and allow the people in my work group to work and make decisions on their own. I occasionally monitor their work to stay informed and make sure needed resources are available. I let the people in my work group set their own pace and determine the ways to accomplish their work assignments. I interact minimally in a straightforward factual way, with little or no daily contact. I ensure that disruptions beyond my work group's control do not occur. I do not make decisions or solve problems for my work group. Sometimes they perceive me as avoiding, withdrawing, permissive or indifferent.

**LIST THE LEADERSHIP SKILLS YOU NEED TO DEVELOP AND WHY.**

## RESOURCES

Amabile, T. (1983) The Social Psychology of Creative Thinking. New York: Springer-Verlag.

Amabile, T. And S. Gryskiewicz. (1987) Creativity In The R&D Laboratory. Technical Report. Center For Creative Leadership, Greensboro, North Carolina.

Glassman, E. (1986) Managing for creative thinking: Back to basics in R&D. R&D Management 16: 176-183.

Glassman, E. (1991) The Creativity Factor: Unlocking the Potential of Your Team. San Diego, CA: Pfeiffer Books of University Associates.

Glassman, E. (1991) For Presidents Only: Unlocking the Creative Potential of Your Management Team. NYC: The Presidents Association of the American Management Association.

Glassman, E. (1996) Creativity Handbook: A Practical Guide To Shift Paradigms And Improve Creative Thinking At Work, a 250 page book written for use in his creativity & innovation meetings & workshops.

Glassman, E. (2010) Team Creativity At Work-I: You Do Want To Be More Successful Than Your Competition, Don't You?

Glassman, E. (2010) Team Creativity At Work-II: Brainstorming Isn't Creative Enough Anymore.

Gordon, T. (1977) Leader Effectiveness Training (L.E.T.).

Hersey, P., and Blanchard, K. (1982) Management of Organizational Behavior: Utilizing Human Resources. (4th ed.) Englewood Cliffs, NJ: Prentice-Hall.

Herzberg (1968) One more time: How do you motivate employees? Harvard Business Review (January, pages 53-62).

---

# • PART 7 •

## CONVERT IDEAS INTO PROFITABLE INNOVATIONS IN R&D

---------------------------------------------------------

### • CHAPTER 28 •
### THREE CORPORATE R&D INNOVATION IMPROVEMENT PROGRAMS

---

The first chapter of this book describes 'Creativity & Innovation Meetings' designed to produce ideas & proposals to solve major problems of a company. The subsequent chapters delineated details of that process. This chapter deals with some ways to **turn ideas into profitable innovations**.

Innovation needs idea-improvement, and **idea-improvement** needs special effort in most companies. People in R&D with ideas may not have the business sense to see its potential or limitations, or the innovation idea may lack data and clarity so no one sees its value when presented to management. An **'Idea-Enhancement Innovation Program'** minimizes these problems by providing R&D idea-people with the means to develop and evaluate their own idea before presentation to management, and by enabling management to make informed decisions about the idea for an innovation.

I discussed this with the head of 'innovation idea-improvement programs' in three Fortune-500 corporations to discover how they succeed. They remain anonymous.

These innovation programs solicit innovation ideas, enhance them, and then persuade management to support the innovation-idea with resources. Such innovation support systems help R&D idea-people to become involved in the identification and early development of ideas for new business opportunities through new technology, new products, and new processes.

In other words, these innovation idea-enhancing programs enable R&D people to dress up their idea before review by senior management who provide the resources for further development.

---

## CORPORATION A

R&D employees submit an   for an innovation to one of the full time 'innovation idea-helpers' at each company site. This helper person works with the idea-proposer to help develop and enhance all technical and business possibilities of the proposed innovation. This service includes market research, patent searches, even some preliminary research. The innovation program then sends the proposal to experts within the company to evaluate and enhance the idea further. The purpose is to overcome snags and "improve the idea and make it work, rather than kill it,"

Then, a team of volunteers from R&D, marketing, and manufacturing work on the innovation idea under a company policy of allowing people to spend 10% time on bootleg activities. Finally, the volunteer team and the innovation idea-proposer publicize the idea, and persuade someone in management to sponsor and provide resources for the idea's development into a commercial product. On average, this innovation process takes about a year.

The program also fosters an innovative environment through a newsletter, bulletin boards, speakers, videos at lunch, and leads creativity sessions when requested to solve a major problem.

## CORPORATION B

The manager of the Office of Innovation told me that his company wants to dramatically speed up the development of new innovation ideas and turn them into new businesses. "We need to develop new products, as well as new uses and new markets for old products," he said.

The Office of Innovation helps the R&D innovation idea-person find seed money, resources, and guidance within the company, frame a presentation, and research the marketplace for the proposed idea. A telephone call starts this informal innovation process, which lacks forms or fixed procedures. The Office of Innovation has substantial funds for research, training, and seed grants and brings in consultants for team creativity training.

Ideas for an innovation without a 'champion,' the person who pushes the idea and turns it into reality, will probably die. So the Office of Innovation encourages people who send in ideas to become idea-champions by helping them perfect their idea. It directs innovation idea-champions to people and resources within and outside of the corporation who will help improve the idea. It brings in business development and market research people to provide guidance on how to develop a business concept proposal.

When well developed, a screening committee decides whether the corporation should provide funds for further development. This committee can fund further development or it can form a business concept team. If the idea turns into a new business, the idea-champion can climb aboard, or return to the old job.

## CORPORATION C

The Center for Creativity & Innovation has three major thrusts. First, to educate people in advanced creative thinking techniques. Second, to apply these creative thinking techniques to important business and technological problems. Third, to help managers create a climate conducive to creativity and innovation.

The Center exposes people to internal and external experts who teach creative thinking techniques; it fosters networking between people interested in creative thinking, and it arranges creative problem solving events that tackle important business and technological problems. This last strategy is essential to impact the bottom line and show the value of creativity & innovation.

I asked the Director: "How did a scientist-administrator get involved with creative thinking and innovation?" He said: "I started reading and attending seminars on creative thinking, and realized there were good resources and workshops outside of the company that would help us be more effective.

"I circulated memos on what I learned, and they stimulated other people. I started a discussion group within the corporation that is now a large network of hundreds of company people in many countries.

"Eventually, the bottom line successes of applying creative thinking techniques to business and technological problems prompted corporate management to ask me to spread creativity & innovation throughout the company..."

Apply these innovation concepts to your business. Make a list of ways to solicit new ideas in your company for new markets for existing products, and new products for old and new markets. Describe how idea champions can be encouraged to develop and pursue ideas in which they have fallen in love. List ways resources can be found and provided to Idea-champions to turn raw ideas into thriving new businesses.

---

## • CHAPTER 29 •

## FRANCHISE INNOVATION TO HELP YOUR R&D TEAM

---

Fostering a sense of ownership among employees can unleash tremendous capacity for innovation. Like franchisees, who own their independent businesses but operate within a well-defined structure or system, empowered employees tend to be more entrepreneurial and committed to helping the organization to thrive.

I had just discussed leading a Creativity & Innovation Meeting with a senior vice president of a large bank who wanted "creative ideas and paradigm shifts" (his description) to help his organization achieve ambitious goals. Going home, my colleague and I acknowledged the restrictive regulation of banks, and focused our creative thinking on internal structuring.

It struck me that in some ways the branches of a bank are like the franchises of a chain of stores. I then realized that all organizations, even teams within them, can be viewed as a hierarchy of empowered franchises.

A franchise is a system of doing business that enables large numbers of independent business people to purchase a ready-made, structured "system" for setting up a store or fast-food establishment, such as the ubiquitous McDonald's chain.

Franchisees tend to be successful because they offer a unique combination of structure and creative freedom. The business person remains an independent entity, and can do as he or she chooses, within certain operational guidelines. Innovative practices within a store can be communicated among other franchisees, raising the level of performance of all stores.

Fostering a sense of ownership is key. People who feel they have a stake in their work are likely to be more committed and creative. The closer you come to this empowered franchise model, the more likely you are to find creativity and innovation, a sense of ownership, the entrepreneurial spirit, personal growth and achievement, a friendly environment, enjoyment at work, cooperation, teamwork, excellent customer relations and self-direction.

---

Each organization is a hierarchy of empowered franchises. Each level in the hierarchy is a franchise that grants an empowering franchise to the lower level. When viewed this way, many options and possibilities open up. For example, I recently read about how a bank empowered some branches to choose their own hours. With this new-found freedom, these branches soon started making other decisions that generated new business and increased profitability.

The type of franchise you grant will depend on the outcome you want. Do you want a sense of ownership, independent thinking, new ways to do things, a creative cauldron, innovation? Then grant a franchise that encourages this. If you want a tighter level of control, you can do that, too (although you will undoubtedly stifle innovation in the process).

This franchise metaphor is useful because there are hundreds of franchises you can analyze for inspiration. Each one empowers the franchisee in slightly different ways, each with its own effect on the creativity and innovation of the employees who work within them.

---

• CHAPTER 30 •

## CREATIVE SMALL BUSINESSES DELIBERATELY
## FOSTER INNOVATION FOR SUCCESS

---

I have interviewed many people and written many articles about how creative small businesses foster innovation in their people. Here are some major strategies they used worth considering for R&D.

### • The Roles of Top Management.

Top management viewed innovation as important to the success of the business, and to remaining competitive. Management deliberately called for innovation. Policies solicit new ideas, reduce bureaucracy, encourage change and different ways of doing things, foster the entrepreneurial spirit, and the belief that people want to be creative. Management tends to give little direction and few guidelines on the implementation of agreed upon goals, respects people's competence, encourages risk and helps people learn from mistakes, wants people to excel and achieve, and promotes from within. They demand practical, profitable results.

### • Power Sharing.

People used words like autonomy, freedom, empowerment, independence, and individual thinking. They're urged to make decisions, create solutions to work problems, and told that the best person to solve a problem is the one working on it. Not a lot of permissions are needed to get the job done with room to innovate. Once there is agreement on the goals, they're given the creative freedom to do the job. Teams are assigned missions, and then turned loose to achieve goals. People are trusted and relied on to make on-the-spot decisions to help customers.

### • Hiring.

These businesses hire diverse people with untraditional backgrounds, good people given lots of leeway, thinking, talented people who are trusted to do the job.

---

- **Rewards.**

Creativity is enjoyable, and that's one reward. Ideas are also rewarded with recognition and full credit. Profit sharing and bonuses were mentioned to motivate people to either implement or tell new ideas to management.

- **Informality.**

People highlighted reduced bureaucracy, vague or no job descriptions, few rules to limit creativity and innovation, fluid organization structure, lack of pigeonholing, no dead end jobs, informal interaction, calling people coworkers (not subordinates), informal job structures, and more.

- **Time.**

Many people mentioned enough time to be creative, and setting deadlines to encourage creative thinking and innovation.

- **Creative Climate.**

Most people used phrases like contagious creativity, friendly environment, be innovative and solve problems creatively, solicit and listen to new ideas, creative physical environment, individualized work area, celebrations, proximity to creative people, caring people, people feel valuable, decent treatment, catch people doing things right, sense of ownership, personal growth, achievement, and self-direction, and more.

- **Teams and Teamwork.**

Most people mentioned teams, cooperation, and creative teamwork. They used phrases like fluid teams, respect each other's competence, high trust, clearly agreed on goals, being open to new ideas, creativity techniques, getting out of the box, and more.

- **Practical Creativity.**

Creativity focused on practical innovative results. People started with spaced out, grandiose ideas, and business realities brought them back to earth.

• **Sources for New Ideas.**

Many people spotlighted outside stimulation to spur innovation. They mentioned colleagues, other people, books, travel, competitors, trade fairs, vendors, magazines, customer suggestions, other stores, and team meetings as sources for their ideas. Ideas are not created in a vacuum. Creativity depends on past experiences and knowledge, so the more you know and interact with others, the more creative and innovative you can be.

• **Sharing Knowledge, Ideas and Values.**

Some businesses train and publish newsletters to keep their people informed. They share new ideas to foster creativity and effectiveness.

Many of these strategies to foster innovation can work in your R&D work group. Make an action plan to introduce those strategies that you think would increase innovation in R&D.

**WHO ...**

**DOES WHAT...**

**WHEN...**

**WHERE...**

**HOW...**

**And WHY...**

> ## • CHAPTER 31 •
> ## INCREASE INNOVATION:
> ## MODERATE THE MID-CAREER SAG IN CREATIVE OUTPUT & PRODUCTIVITY

In 1953, Herbert Lehman published "Age and Achievement" in which he suggested that scientific achievement peaked between 30 and 40 years of age, and declined after that. Many studies confirm this theory. One such study is reported in Chapter 10 of "Scientists In Organizations," by D. C. Pelz and F. M. Andrews published in 1976. Their results are very pertinent today.

Over 1300 scientists and engineers were studied: 641 in 5 companies (electronics, electrical equipment, glass, ceramics, and pharmaceuticals); 526 in 5 government laboratories (weapons guidance, animal diseases, agricultural products, and physical sciences); and 144 faculty in 7 departments of a major university.

Productivity (and innovation) was evaluated in the following ways: professional papers published and patents obtained in the past five years; unpublished reports circulated within the company in the past five years; overall usefulness to the organization and contributions to specialized field of knowledge rated by colleagues.

The results are startling. All groups showed a peak in productivity at about 45 years of age, much lower productivity at age 50, and a small rise in productivity at about age 55. The findings on engineers in development laboratories are of special interest since they resemble most technical groups found in industry. Their productivity (as measured by contributions and overall usefulness to the organization) was highest between ages 45 and 49. Productivity was much lower between ages 50 to 54, and it was slightly higher in ages 55 and older.

What could account for this **mid-career productivity sag** followed by a mild recovery? Self-reliance or a desire for self-direction seemed important. Engineers who were self-motivated to work on their own ideas, who desired more freedom, and who were curious and stimulated by their work had higher levels of productivity at ages 45 to 49 than those with weak self-motivation and weak desire to work on their own ideas. Most interesting, the productivity sag seen past age 50 was not as great in engineers who showed self-reliance and desired self-direction, and their recovery over age 55 was stronger. Thus, strong self-motivation seems important to reducing the mid-career productivity sag.

## How to moderate the mid-career productivity sag and lessen its impact?

Some general approaches are useful.

First: Accept it as normal. Look for it. Talk about it in a supportive way. If handled well, it is not the end of a fruitful career or the start of a permanent decline.

Second: Help younger people prepare for the productivity sag with early career planning to reduce or prevent it. Plan for renewal at appropriate times using sabbaticals, job rotations, courses, training, advanced degrees, etc.

Third: Consider ways to revive self-motivation and involvement in older people. Give them choices that include new projects, transfers, and additional training. Allow them to become intrapreneurs or to move into consultant or administrative roles. Train them to be mentors and supportive helpers of younger people.

Fourth: Assume the productivity sag is not inevitable or irreversible. The effects of age can be modified. Special training in self-reliance and assertiveness may help people want to be more self-directed.

Older employees may represent a valuable untapped human resource for companies that a creative approach may unleash. Does anyone want to arrange a creativity meeting to solve this problem?

---

# • PART 8 •

## APPENDICES

---

### • APPENDIX I •

### CREATIVE THINKING TECHNIQUES

"When you come up with the obvious, look elsewhere."

---

People find advanced creative thinking techniques indispensable when working alone or in teams that want to solve important problems.

Techniques marked with an * indicates one of the key creative (non-evaluative) steps.

**\*Step 1.**
Define the problem creatively.
    Essence or action verb
    Like-improve analysis
    List how-to statements
    The problem's essence
    Reversal-dereversal
    Who, what, where, when, and why
    Word substitution

**Step 2.**
Identify the criteria to select the right problem.

**Step 3.**
Select one or more reasonable problems.

**\* Step 4.**
List many ideas and solutions.
    Automatic writing
    Bizarre trigger-ideas
    Brainstorming
    Brainwriting circles
    Buzz groups
    Clustering brainwriting
    Combining ideas team
    Dream interruption brainwriting
    Forced-combinations
    Idea card

    Idea gallery
    Improve bizarre trigger-idea game
    Metaphors
    Non-evaluative listing
    Set a quota for new ideas
    Trigger-ideas
    Weird to workable idea

**\*Step 5.**
Combine ideas into trigger-proposals innovatively.
    Forced-withdrawal
    Idea board
    New-company
    Return-to-reality
    Trigger-proposals

**Step 6.**
Identify the criteria to select appropriate proposals.

**Step 7.**
Generate workable solutions and action plans.

**Step 8.**
Sell proposals resourcefully.

**Step 9.**
Innovation Improvement Program

**EDWARD GLASSMAN, THE AUTHOR, IN HIS 80TH YEAR**

www.ingramcontent.com/pod-product-compliance
Lightning Source LLC
Chambersburg PA
CBHW081114170526
45165CB00008B/2447